P ROJECT

APOLLO

The Test Program

by Robert Godwin

All rights reserved under article two of the Berne Copyright Convention (1971).
We acknowledge the financial support of the Government of Canada through the Book
Publishing Industry Development Program for our publishing activities.

Published by Apogee Books, Box 62034, Burlington,
Ontario, Canada, L7R 4K2, http://www.apogeebooks.com
Tel: 905 637 5737
Printed and bound in Canada
Project Apollo The Test Program by Robert Godwin
ISBN 1-894959-36-1
©2005 Robert Godwin

The Apollo Lunar Test Program
1964-1969

Between August of 1963 and July of 1969, when men finally flew an Apollo spacecraft successfully to a lunar landing, the hard-working NASA employees diligently pursued an exhaustive regime of test flights. Beginning with the unmanned test flight of the Little Joe QTV (which tested an Apollo shaped structure in flight for the first time) and concluding with the final skimming of the Lunar surface by the crew of Apollo 10, the Apollo Saturn test program stands as one of the greatest engineering achievements in the history of our species. This is not mere hyperbole. The scope of this effort has often been compared to the building of the great pyramids or the Manhattan Project. What is truly remarkable is that the landing on the moon was the culmination of a mere six years of test flights. Admittedly the mighty Saturn rockets and their enormously powerful engines had a heritage that could arguably be traced back to the end of the 19th century, but it was only with the announcement made by President Kennedy in May of 1961, that Americans would go to the moon, that Apollo was conceived. It would be just 36 months from that momentous speech to the first foray into space by an unmanned lunar spacecraft. Such schedules today seem almost inconceivable and are a testament to the youthful spirit of an American people; invigorated by their charismatic President with the promise of a better future—a spacefaring future.

Most histories of the Apollo program gloss over the incredible string of events that comprised the Apollo test program. It is rare to find any mention made of the flights before Apollo 7, which was the first Apollo to carry humans into space. Occasionally people wonder about Apollo 1 through 6 (making the assumption that if there was a number seven there must have been numbers one through six) but the average citizen is hard pressed to find a concise overview of these extremely important missions. This book is an attempt to fill that void.

The numbering systems that NASA officials applied to the test program are as impenetrable as the acronyms and jargon espoused during the golden era of space flight. There was no Apollo 1, at least not at the beginning.

Before an Apollo could fly in space there had to be several test flights of the Saturn rocket. The first iteration of Wernher von Braun's giant booster was called the Saturn I (One). This huge rocket stood over 160 feet tall and delivered an unprecedented 1.5 million pounds of thrust. On October 27th 1961 (just five months after Kennedy's speech) the SA-1 roared to life and carved an arc across the Atlantic sky. It was a sub-orbital flight that took the rocket 85 miles up and 207 miles downrange. The payload was an aerodynamic fairing which contained nothing. Six months later, on April 26th 1962 the SA-2 carried 96 tons of water in its upper stages which were deliberately exploded 65 miles above the earth. It was an experiment called Project High Water and, in an effort to get something more out of the launch than just testing the rocket, an artificial cloud was created at high altitude for research purposes. Then on 16th November 1962 an almost identical experiment was conducted with the launch of SA-3.

SA-4 was launched on the 28th March 1963 and was slated to be the last sub-orbital test of the Saturn I rocket. The primary goal was to shut down one engine deliberately early and see if the vehicle could continue on its assigned trajectory. The mission was a total success.

Later that summer, on August 28th, a cone-topped cylinder that was just over 23 feet high and almost 13 feet in diameter was launched on top of a small rocket 33 feet high. It was an empty shell but it had the basic aerodynamics of NASA's proposed lunar spacecraft. The rocket was known as "Little Joe" and it used seven engines, one large engine known as an *Algol* and six smaller engines known as *Recruit*. The main purpose of this flight was to see if the "Little Joe" launcher was capable of launching a real Apollo command and service module. The mission was mostly a success and the empty replica flew to an altitude of 5.6 kilometers. Just over ten weeks later the first of two "Pad Abort Tests" took place at White Sands Missile Range. This version of the spacecraft looked outwardly much the same as its predecessor but it was what was known as a "boilerplate", a dummy of the final spacecraft that was still taking shape at the North American Aviation factories in Southern California. It was given the somewhat inglorious name of Boilerplate 6 (BP-6) and it was launched without the benefit of a "Little Joe" booster. The only thing carrying the BP-6 capsule would be a thirty foot tall scaffold known as the Launch Escape System (LES), a tower structure which sat atop the Apollo command module. The LES was designed to yank the command module away from any potential failure during launch, using an array of very high powered solid fuel rockets. The PA-1 test took place on November 7th 1963 and was a complete success, lifting the BP-6 to an altitude of 2.8 kilometers.

Almost three months went by before the next Saturn launch took place, it was designated SA-5 and this time it would be a full orbital test. The rocket would be the first Block 2 version of the Saturn rocket and it would take a fully fueled upper stage, called the S-IV, into orbit where the liquid hydrogen/oxygen engines would be tested. This was the first test of the proto-type stage that would be necessary to later take men to the moon. This early version would have six RL-10 hydrogen engines. Almost 38,000 pounds were delivered to orbit by SA-5.

On 13th May 1965 another "Little Joe" was prepared for launch on Pad 36 at the White Sands Missile Range in New Mexico. This one was carrying Boilerplate 12 (BP-12). This prototype spacecraft was loaded with measuring equipment and flew to an altitude of almost six kilometers.

Up until this time, nothing which even remotely resembled an Apollo spacecraft had been into space. This would finally happen with the launch of SA-6 on 28th May 1964. It would not be a manned mission, it would not even be an operational spacecraft, it was another boilerplate (BP-13) and it was placed into an elliptical orbit by the new S-IV stage, which burned its six high thrust engines for 473 seconds. During the flight one engine shut down early but the mission proceeded unimpaired. Apollo was finally in space, albeit only a clunky facsimile of what was to come.

With the success of SA-6 another Apollo boilerplate vehicle was prepared for launch, and on 18th September 1964 SA-7 took an S-IV stage with a dummy Apollo command and service module into an 88 minute orbit around the earth. Over 130 measurements were made by the spacecraft and launch vehicle and finally proved that the Saturn rocket was ready for prime time. The boilerplate Apollo had the same size, weight, shape and center of gravity as the real thing and the flight was so successful that the next three missions, which had originally been slated as further test flights, were altered to become science missions.

Just about three months later a third "Little Joe" was launched at White Sands, this time with two Algol engines supported by four Recruits. It carried BP-23 and its primary goal was to test the Launch Escape System (LES) at altitude while the vehicle was undergoing high stresses as it passed through maximum aerodynamic pressure. The launch happened on 8th December 1964 and almost immediately revealed a design problem with a protective shroud called the Boost Protective Cover (BPC). This thin layer of laminate and cork, designed to protect the command module, was partially ripped away but the BP-23 returned safely to the ground. A fourth Little Joe flight took place on May 19th 1965 (A003) using six of the high power Algol engines. The capsule was BP-22 and the launch was a complete failure, which ironically proved the worth of the LES. The first stage of the Little Joe failed within 3 seconds of lift off and broke apart. The BP-22 was ripped away by the LES at low altitude and proved that it was possible to survive a failure at only 12,000 feet.

The Apollo Saturn combination had been declared operational after the flight of SA-7. What followed next were (in this order) SA-9, SA-8 and SA-10. The flight numbering sequence was interrupted by the decision to start using contractor-built stages, so SA-9 was the last Saturn I built by NASA in Huntsville and SA-8 was the first built by Chrysler. Chrysler were running a little late and so SA-9 hit the pad on February 16th 1965 followed by SA-8 on 25th May and finally SA-10 on July 30th. All three launches carried large satellites called Pegasus. These were stowed inside the boilerplate Apollo service module and the third stage, and were designed to study the dangers of micro-meteoroid impacts in earth orbit. The Pegasus were built in Huntsville by the Saturn team and all three were placed in relatively high orbits (above 300 miles). This was the end of the Saturn I series of rocket launches.

On June 29th 1965 another Pad Abort Test had taken place at White Sands this time using the refurbished BP-23 capsule (renamed BP23A). This second LES-only test successfully cleared away most of the problems with the BPC and with the steering capabilities of the LES. After the success of SA-10's flight, one more LES launch took place at White Sands. This was another full-up Little Joe using 4 Algol engines with five Recruits. On January 20th 1966 Little Joe II A004 took flight to an altitude of over 20 km where a deliberate abort ripped the booster apart concluding the Little Joe/LES/PA test series.

The next generation of Saturn rockets was now ready for trials. It was to be called the Saturn IB (S-IB) and it was designed to do some early testing of lunar hardware, while the giant Saturn V lunar rocket was still under construction. The S-IB used a very similar first stage to the S-I, although it had a different fin structure at the bottom, and the top was altered to accommodate a larger upper stage called the Saturn IVB (S-IVB). On 26th February 1966 the first S-IB rocket lobbed the first true Apollo spacecraft on a sub-orbital trajectory. It was named after the launch vehicle (SA-201) and so this first mission became Apollo/Saturn 201 (AS-201). Using an imaginative program of engine burns the AS-201 Apollo command module was sent into a high energy re-entry path to test out the heat-shield of the spacecraft. The mission was a success and the Apollo command module splashed down successfully just 32 minutes after launch. Shortly after this flight, on May 25th 1966 an almost complete Saturn V was rolled out to the #39 launch pad. It would not fly as it was strictly a test vehicle numbered AS-500F.

The next flight would be an orbital test of the new S-IB rocket and the first full orbital flight of the upper lunar stage, the S-IVB. It was called AS-203 and was launched 5th July 1966. No Apollo spacecraft was attached to this booster as the primary goal was to investigate the behavior of the fuel inside the S-IVB. Instead, an aerodynamic fairing was placed on top. The mission revealed many new facts about the launch vehicle and was only marred by an over-pressurization of the S-IVB during its fourth orbit, which led to the stage breaking up.

The wheels at NASA were definitely in full gear allowing the next Saturn launch to take place only seven weeks later, on 25th August. This would be the first time that an Apollo command module would be subjected to the full force of a high speed re-entry. The AS-202 command module would only be sent into a suborbital trajectory but it would re-enter at a punishing five and a half miles per second, almost the speed that a returning lunar vehicle would attain. The delicate cone-shaped craft would be lofted up over 600 miles high by using the S-IVB engines as well as the new Service Module engine. About an hour and a half after launch, and travelling over 17,000 miles downrange, it splashed down successfully in the Pacific ocean.

On the books at this time were plans for more "AS-2" missions, at least all the way through to AS-209 but they were never to fly. The next vehicle to be erected on the launch pad at Cape Canaveral was the ill-fated AS-204.

Delayed by a raft of problems, AS-204 was installed atop a Saturn IB launch vehicle on Pad 34. It was one of the early Apollo capsule designs, the so-called Block I vehicle. AS-204 was to have been the first manned flight of an Apollo spacecraft. The crew was led by one of NASA's top veteran astronauts, Virgil "Gus" Grissom and he was accompanied by America's hero space-walker, Edward White, and Roger Chaffee, a navy lieutenant commander with no prior space-flight experience. The AS-204 mission had been subjected to a host of delays caused by the vehicle's environmental control system, and further exacerbated by a fuel tank that had exploded at the contractor's factory, completely destroying an Apollo Service Module. By the time the cause of the explosion had been determined the Apollo program was slipping behind schedule and the problems were being compounded by funding deficiencies. Losing a Service Module was an expensive blow, but such engineering challenges were to be expected. It was, however, an unheeded warning of much worse to come.

At precisely 6:31:04, eastern standard time, on the 27th January 1967 during a routine systems check on the launch pad, a fire was noticed by the crew at the front left interior of their AS-204 spacecraft. Within eight seconds it had burst into a ferocious intensity, fueled by the cabin atmosphere that was composed of 100% oxygen at high pressure. Seven seconds later the pressure inside the cabin was at 36 pounds per square inch and the spacecraft floor ruptured on the right-hand side of the capsule. Flame shot up and enveloped the outside of the spacecraft effectively preventing the ground technicians from providing assistance. The crew were almost certainly overcome by smoke within seconds since they couldn't open the hatch, which opened inwards and was held in place momentarily by the enormous internal pressure. There was nothing any of the dedicated NASA team could do, it was too late, and three brave men gave all they had in their dream to reach the moon. In a belated act of uncharacteristic clarity NASA management redesignated the ill-fated mission-Apollo 1.

After an intensive investigation into the causes of the fire it was determined that the manned Apollo program would have to await the arrival of the overhauled Block 2 spacecraft. This newly designed vehicle was to have flown, unmanned, as part of missions AS-205 (which had its goals combined with 208 to save time) and AS-207 (which was coupled with 209). Now in the wake of the loss of Apollo I everything in the schedule was overturned. The next flight would finally be given a designation that attempted to simplify the mission naming game, although presumably someone must have made the decision to deliberately leave out the boilerplate flights and the non-Apollo AS-203 and to only count the two authentic Apollo spacecraft that had flown (AS-201 and 202) along with the tragic Apollo I mission. This mission was to be called Apollo 4 and it was to be the first all-up test of the mammoth Saturn V rocket.

It was evidently not necessary to use the new Block 2 Apollo spacecraft for this particular trial, since no lives were at risk, and so a slightly modified Block I was hurled aloft by the immense propulsion of Von Braun's giant rocket. Nestled inside the fairings just below the Block I command and service modules was an ungainly chunk of hardware which had been placed there as ballast. It was a bizarre contraption that could scarcely be called anything but ugly but it was only an embryo of what was to be spawned at the Grumman factory in Long Island, New York. This misshapen lump of metal and wiring was called the Lunar Module Test Article (LTA) and in keeping with NASA's penchant for incomprehensible numeracy it was designated number 10. There are no pictures of the LTA flying in space, which is a shame, but its purpose was merely to monitor the environment inside the snug confines of the third stage (S-IVB) fairing, called the Spacecraft Lunar Module Adapter (SLA). Its even more ungainly but vastly more complicated sibling, the Lunar Module, would have to sit in the exact same place and undergo the exact same jarring vibrations during launch. The LTA is the lost cousin of the Apollo program, the unmanned spacecraft which flew and nobody noticed—no one except the engineers at Grumman who were struggling to shed every molecule of weight from their greatest creation; a conglomeration of wiring, pipes, glass and metal which had to comfortably sustain two men while they lived on another world.

After the earth-shaking debut of the Saturn V and the successful flight of Apollo 4 (which even though it was a Block I capsule tested the new Block 2 heat shield during re-entry) Grumman's intricate and bug-like lunar module was ready to take centre stage. It would be launched atop another Saturn IB this time without command and service modules. The first LM was almost exactly identical to the machine slated to go to the moon except it was missing its landing gear. The top of the second stage of the S-IB would be wrapped in an aerodynamic fairing and nose cone, to protect the LM, the mission would have two designations, Apollo 5 and SA-204. (note that's SA, not AS.) Almost exactly a year after the terrible launch pad fire the night sky at the Cape was flushed by the glow of the giant H-1 rockets roaring to life as Grumman's masterpiece finally sallied forth to its natural environment. The controllers on the ground pushed the first true spacecraft through its paces by remote control. Alternately firing and shutting down the engines which would one day take people to another world.

Just over ten weeks later the second flight-worthy Saturn V was rolled onto the launch pad. It was carrying yet another Block 1 command and service module and another of the ugly lunar module step-sisters, the Lunar Module Test Article, this one was number 2...

It is perhaps worth noting that there were in fact dozens of pieces of hardware scurrying around the country that comprised "bits" of Lunar Modules. A half dozen mockups, ten test models, sixteen "Ground Program Articles" (including three LTAs which flew), at least three simulators and then no less than sixteen actual lunar modules. Numbering ran M1 through 6 for the mockups, TM-1 through 9 for the test models and LTA 1 through 10 for the Ground Program Articles (with a bunch of PDs and PAs thrown in for good measure). The average journalist couldn't possibly hope to convey the complexity of this seemingly illogical naming system and so none tried. It was simply easier to give up and start calling the missions Apollo—followed by a number.

The Apollo that raced into space in April of 1968 was Apollo 6 and it would successfully clear away many more problems presented by the launch of such an incredibly complicated machine. The LTA (2R) was marginally modified to make allowances for data collected by Apollo 4's LTA. The mission objectives (outlined further on in this book) were mostly accomplished and the stage was finally set for another shot at a manned mission, this time on the ground.

Between May 27th and November 14th 1968 extensive testing was conducted inside the huge vacuum chambers at the Manned

Spacecraft Centre in Houston. These mostly forgotten missions consisted of putting ten pilot/astronauts inside the newly revised Block 2 spacecraft and inside one of the latest iterations of the Lunar Module Test Articles. The two spacecraft were Command Module 2TV-1 and Lunar Module LTA-8. The crews for the Command module missions included Joe Engle (who would eventually command Shuttle missions), Vance Brand (who would fly on Apollo-Soyuz) and Joe Kerwin (who would fly to Skylab.) and Air Force Majors Turnage Lindsey, Lloyd Reeder and Alfred Davidson. On the LM side it was Jim Irwin (who went to the moon on Apollo 15) and Grumman test pilots Gerald Gibbons, Glennon Kingsley and Joseph Gagliano. These crews were given the task of simulating an Apollo flight with all of the attendant dangers of space, while never leaving the ground. The chambers were sucked clean of air and huge solar lamps heated up the spacecraft to broiling temperatures while packs of liquid nitrogen were used to cool them down. At one point the 2TV-1 crew even conducted a fake space walk in full pressure suits. Even though they never left the ground, these crews faced very real danger putting these experimental spacecraft through their paces. These missions were a resounding success and cleared the way for finally putting Apollo into space with a live crew aboard.

The follow-on crew from Apollo 1 had been training vigorously for nearly two years since the fire and the designated commander, Wally Schirra, was on top of as many aspects of the newly designed Block 2 spacecraft as was humanly possible. Schirra was in a no-nonsense mood and he carried the responsibility of returning his country to space in as dead serious a fashion as his normally gregarious character would allow. Apollo 7 was to spend the equivalent time of a lunar mission in low earth orbit; just as Apollo 1 had been slated to do. Up to fourteen days was the mandate, but this time in a completely revamped spacecraft. Schirra and his crew were ready to take the bull by the horns and show that America was back on track. Since no Lunar Module was to fly with Apollo 7 the fairing of the second stage contained only a framework and docking target so that CM pilot Donn Eisele could fly an approach that would simulate the docking necessary for later missions. After launch the crew of Apollo 7 found themselves slipping into a relatively uninteresting regime which, combined with Schirra's head-cold, led to some frayed tempers. However, the mission was a splendid success. America's three-man spacecraft was evidently ready.

Few people monitoring the program were prepared for the announcement which seeped out of the Apollo program office short-ly after the splashdown of Apollo 7. Rumors abounded that the late Soviet chief designer Sergei Korolev's giant lunar rocket, the N-1, was slumbering near a launch pad on the Kazakh steppes. Tension mount-ed with the prospect that the Soviets might upstage the methodical American agenda and so, after much consultation, it was decided that the next flight would risk double jeopardy. Apollo 8 would not only carry humans atop a Saturn V, for the first time, but it was to go to the moon.

In August of 1968 the Saturn program management had expressed some concerns about the weight of the Apollo 8 payload. Grumman were still behind and none of the delicate spidery landers were ready to fly. It was decided to include another Lunar Module Test Article inside the SLA petals of the S-IVB stage. This would again act as ten tons of ballast, simulating the presence of a lunar module. The LTA-B would never leave the confines of the S-IVB. The overall shape of this one was different to the earlier LTA's that had flown on Apollos 4 and 6. LTA-B was basically a hollow cylinder with the vestiges of LM legs. Crews were shifted in the flight order and the highly trained LM crew were shunted to Apollo 9 while Frank Borman, Jim Lovell and Bill Anders discovered that they would be spending their Christmas 60 miles above another world. On December 21st 1968 at 7:51 EST the first mission to the moon left Florida roaring its way atop a column of fire spewed forth by five giant Rocketdyne F-1 engines. It was a spectacle of biblical proportions as the engines consumed the fifteen tons of propellant every second necessary to lift the six million pound monster into space. The restartable, third stage J-2 engine dili-gently performed its prescribed function and sent the three men into cislunar space.

By the time Apollo 8 completed its goals and the three intrepid explorers dragged themselves into their rubber rescue dinghy in the Pacific ocean, the Apollo program was running critically late. The lunar landing vehicle had yet to fly with people in space, and the first vehi-cle had only arrived at the Cape the previous June. NASA crews had a mere nine months to prep the prickly beast and get it ready for flight. President Kennedy's timeline was a man on the moon and safe-ly home by the end of the decade—only twelve months remained.

The Saturn V had lived up to and exceeded all hopes and expectations and the management team were confident that they were ready to finally fly the last piece of the giant Apollo puzzle. It meant placing two people inside a spacecraft with walls that were absurdly thin and having them fly away from the much more robust command module. This would only be the second time that NASA had tried to control two manned spacecraft independently and simultaneously. Despite these concerns Apollo 9 went ahead on schedule, taking Jim McDivitt, Rusty Schweickart and Dave Scott into low earth orbit on March 3rd 1969. The aptly named "Spider" LM was put through its paces and flew independently of its sister command module. The crew took a spacewalk for the first time during an Apollo mission and the subsequent docking was conducted by McDivitt and Schweickart, a precautionary learning procedure in case a command module pilot was somehow incapacitated after future crews returned from the moon. All that remained now was to test the entire set of hardware at lunar distance.

Now, with only seven months left in Kennedy's arbitrarily chosen schedule, the crew of Apollo 10 would fly the final dress rehearsal. On May 18th 1969 Tom Stafford and his all veteran crew of John Young and Gene Cernan, left for the moon. They would take Lunar Module number 4 down to within 47,000 feet of the lunar surface (it was the third to fly but apparently no one was counting). An entire wealth of new information was accumulated and (as you will see in the following text) most of the missions goals were accomplished. The flight of Apollo 10 brought to an end the era of the Apollo test program. From that point forward it could be argued that Apollo was now a fully operational system of hardware. Not all of the bugs were worked out, as the crew of Apollo 13 were to learn, but Apollo would go on to take twelve humans to the surface of the moon and successfully bring them home safely.

Many people on the inside of NASA were already aware of the rumbling noises coming from Congress that would ultimately condemn four flight-ready Lunar Modules to become museum pieces, while one vanished completely, and another ended up in a wreckers' yard (to be plundered and picked clean for precious metals). Even more astonishingly several flight-worthy Saturn V's would be dispatched from the factories only to become outsized lawn ornaments. Before Apollo 11 even left the earth the forces were arraying in Washington to make sure that Apollo would be remembered almost as a daydream.

APOLLO/SATURN TEST FLIGHTS

Mission	Date	Vehicle	Payload	Description
SA-6	5/28/64	SA-6	BP-13	Boilerplate Apollo spacecraft. One inboard engine unexpectedly shut down 26 sec. early but did not impair flight.
SA-7	9/18/64	SA-7	BP-15	Boilerplate Apollo spacecraft. 39,000 lbs into orbit. Declared operational three flights early.
SA-9	2/16/65	SA-9	BP-16 Pegasus 1	Boilerplate Apollo command module. First operational flight. Pegasus 1 inside Service Module placed into orbit.
SA-8	5/25/65	SA-8	BP-26 Pegasus 2	Boilerplate Apollo command module. Pegasus 2 launched inside Service Module placed into orbit.
SA-10	7/30/65	SA-10	BP-9 Pegasus 3	Boilerplate Apollo command module. Pegasus 3 launched inside Service Module placed into orbit.
AS-201	2/26/66	SA-201	CSM-009	Launch vehicle and CSM development. Test of CSM subsystems and of the spare vehicle. Demonstration of reentry adequacy of the CM at earth orbital conditions.
AS-203	7/5/66	SA-203	LH2 in S-IVB	Launch vehicle development. Demo of control of LH2 by continuous venting in orbit.
AS-202	8/25/66	SA-202	CSM-011	Launch vehicle and CSM development. Test of CSM subsystems and structural integrity. Demo of propulsion and entry control by G&N system. Re-entry at 28,500 fps.
APOLLO 4	11/9/67	SA-501	CSM-017 LTA-10R	Launch vehicle and spacecraft development. Demonstration of Saturn V performance and of CM entry at lunar return velocity.
APOLLO 5	1/22/68	SA-204	LM-1 SLA-7	LM development. Verified operation of ascent and descent propulsion and structures. Evaluation of LM staging and S-IVB/IU orbital performance.
APOLLO 6	4/4/68	SA-502	CM-020 SM-014 LTA-2R	Launch vehicle and spacecraft development. Demonstration of Saturn V Launch Vehicle performance.
APOLLO 7	10/11/68	SA-205	CM-101 SM-101	Manned CSM operations. Duration 10 days 20 hours. SLA-5
APOLLO 8	12/21/68	SA-503	CM-103 SM-103 LTA-B	Lunar orbital mission. Ten lunar orbits. Mission duration 6 days 3 hours. Manned CSM operations
APOLLO 9	3/3/69	SA-504	CM-104 SM-104 LM-3	Earth orbital mission. Manned CSM/LM operations. Duration 10 days 1 hour.
APOLLO 10	5/18/69	SA-505	CM-106 SM-106 LM-4	Lunar orbital mission. Manned CSM/LM operations. Evaluation of LM performance in cislunar and lunar environment, following lunar landing profile. Mission duration 8 days.

SATURN/APOLLO (SA) 06
FLIGHT SUMMARY

Saturn launch vehicle SA-6, the second of the Saturn I Block II vehicles, was launched at 1207 hours EST on May 28, 1964. The flight test was a complete success with all missions being achieved. The only significant deviation from expected performance was the premature cutoff of S-I engine 8 after 116.88 seconds of flight.

SA-6

SA-6 was the second vehicle launched from Complex 37B at Cape Kennedy and represented the first launch of a Saturn/ Apollo configuration. The successful launch countdown, concluded on May 28, 1964, was interrupted by four holds. The actual flight path of SA-6 deviated considerably from nominal. At S-IV cutoff the actual altitude was 2.4 km lower than nominal and the range was 41.4 km shorter than nominal.

The S-I propulsion system (eight H-1 engines) performed satisfactorily, but slightly lower than predicted, until 116.88 sec of flight. At this time engine position 8 unexpectedly cutoff. The performance of the S-IV propulsion system (six RL10 A-3 engines) was within design limits throughout the SA-6 flight.

The overall performance of the SA-6 guidance and control system was satisfactory. Separation was executed smoothly; the resulting control deviations were small and easily controlled out.

The ejection of the command module launch escape system tower at separation plus 12 sec had virtually no effect on the vehicle control. Guidance and control system hardware environments were within the specified limits.

The vibration levels experienced on SA-6 were, in general, within the expected limits. The structural analysis revealed that no pogo instability existed during the flight. The vehicle electrical systems performed satisfactorily.

The boilerplate Apollo Spacecraft (BP-13) performance was highly satisfactory with all spacecraft mission test objectives being fulfilled by the time of orbital insertion and additional data were obtained by telemetry through the Manned Space Flight Network until the end of effective battery life in the fourth orbital pass.

TEST OBJECTIVES
(All Accomplished)

1. Launch Vehicle Propulsion, Structural and Control Flight Test with Boilerplate Apollo Payload

2. Second Live Test of S-IV Stage
3. Second Flight Test of Instrument Unit
4. Demonstrate Physical Compatibility of Launch Vehicle and the First Apollo Boilerplate under Preflight, Launch and Flight Conditions
5. First Active Guidance System (ST-124 at Separation + 14 sec)
6. First Test of Guidance Velocity Cutoff (S-IV Stage)
7. S-I/S-IV Separation
8. Second Launch From Complex 37B
9. Planned, Large Angle-of-Attack During High Q
10. Recovery of 8 Movie Cameras which view LOX Sloshing, Separation, Chilldown, etc.
11. Flight Control Utilization of S-I Stage Fins
12. Demonstrate Launch Escape System (LES) Under Flight Conditions
13. Vented Hydrogen During Chilldown
14. Second Launch of Vehicle from Eight Fixed Launcher Arms
15. Separation Initiated by Timer
16. Second Orbital Flight and First Orbital Flight of Burned Out S-IV Stage, Instrument Unit and Apollo Boilerplate; approximate weight 17,100 kg (37,700 lb)

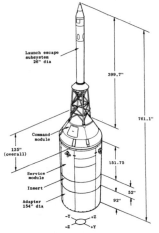

Launch escape subsystem 26" dia

399.7"

761.1"

Command module

135" (overall)

Service module

Insert

151.75

52"

92"

Adapter 154" dia

-Y   +Z

-Z   +Y

APOLLO BP-13 SPACECRAFT (SA-006)

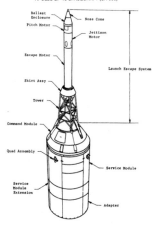

Ballast Enclosure

Nose Cone

Pitch Motor

Jettison Motor

Escape Motor

Launch Escape System

Skirt Assy

Tower

Command Module

Quad Assembly

Service Module

Service Module Extension

Adapter

APOLLO BP-15 SPACECRAFT (SA-007)

SATURN/APOLLO (SA) 07
FLIGHT SUMMARY

Saturn launch vehicle SA-7, third of the Block 11 vehicles, was launched at 11:22 AM EST on September 18, 1964. The flight test was a complete success with all missions being achieved.

SA-7 was the third Saturn vehicle launched from Complex 37B at Cape Kennedy and represents the second launch of a Saturn/Apollo configuration.

The actual flight path of SA-7 deviated from nominal due to high S-I stage performance.

The performance of both the S-I and S-IV stage propulsion systems was satisfactory for the SA-7 flight test. SA-7 was the third Saturn vehicle to employ H-1 engines at a thrust level of 836,000 N (188,000 lbf) to provide thrust for the S-I stage.

SA-7

SA-7 also represented the third Saturn flight test of the RL10A-3 engine for the S-IV stage. The performance of all S-IV subsystems was as expected for the flight test.

The Overall performance of the SA-7 Guidance and Control System was satisfactory. The ST-124 system, along with control rate gyros, provided attitude and rate control for both stages.

Separation was executed smoothly with small control deviations experienced in the pitch and yaw plane. A larger than expected ullage rocket misalignment produced a significant roll deviation of 6.0 degrees.

The vibration levels on the S-I stage were among the lowest ever exhibited by the Saturn vehicle. The S-IV vibrations were about the same as previously observed.

Surface pressures and temperatures on the S-I and S-IV stages were in good agreement with past results. The S-I and Instrument Unit electrical systems operated satisfactorily during the boost and orbital phase of flight. All mission requirements were met.

Recovery of the 8 onboard cameras was impossible because of Hurricane Gladys. Two cameras were subsequently recovered after having been washed up on the beaches at San Salvador and Eleuthera Islands. The Boilerplate Apollo Spacecraft (BP-15) performance was highly satisfactory with all spacecraft mission test objectives being fulfilled by the time of orbital insertion and additional data were obtained by telemetry through the Manned Space Flight Network until the end of effective battery life during the fourth orbital pass.

TEST OBJECTIVES

(All Accomplished except where noted)

1. Launch Vehicle Propulsion, Structural, Guidance and Control Flight Test with Boilerplate Apollo Payload
2. First Complete Flight Test (Both Stages) Utilization of the ST-124 Platform System
3. Second Flight to Demonstrate the Closed Loop Performance of the Path Guidance Scheme during S-IV burn using the ST-124 Guidance System
4. Third Live Test of S-IV Stage
5. Third Flight Test of Instrument Unit
6. Demonstrate Physical Compatibility of Launch Vehicle and the Second Apollo Boilerplate under Preflight, Launch and Flight Conditions
7. Second Test of Guidance Velocity Cutoff (S-IV Stage)
8. Third Test of S-I/S-IV Separation
9. Third Launch From Complex 37B
10. First Flight of Active ASC-15 Time Tilt Polynomial for S-I
11. First Complete Flight Test (Both Stages) Using Control Rate Gyros in Closed Loop
12. First Flight Test Demonstration of the Spacecraft's Alternate LES Tower Jettison Mode Utilizing the Launch Escape Motor and Pitch Control Motor
13. First Test of the S-IV Stage Non-Propulsive Venting System
14. First Test of S-I Engine Area Fire Detection System (Passenger Only)
15. First Test Without S-IV LOX Tank Backup Pressurization System
16. Recovery of 8 Movie Cameras Which View LOX Sloshing, Separation, Chilldown, etc - Not Achieved * Two cameras were subsequently recovered after having been washed up on the beaches at San Salvador and Eleuthera Islands.
17. Third Orbital Flight of Burned Out S-IV Stage and Instrument Unit; Second Orbital Flight of Burned Out S-IV Stage, Instrument Unit and Apollo Boilerplate; Approximate Weight 17,700 kg (39, 100 lbm)

Access Door
LH2 Tank Dome
Manhole Cover
Cold Helium Sphere (3)
Tunnel
Aft Skirt
Ullage Rocket (4)
Umbilical Panel
LH2 Makeup Sphere
Helium Heater and Ambient Sphere
Aft Interstage
Engines (6)
Blowout Panel (8)

Telemetry Antenna (4)
Command Destruct Antenna (4)
Forward Interstage
Cylindrical LH2 Tank
Common Bulkhead
LOX Tank baffle
Aft Bulkhead
Thrust Structure
LH2 Suction Line (TYP.)
Heat Shield
Hydrogen Vent Stack (3)

S-IV STAGE

SATURN/APOLLO (SA) 09
FLIGHT SUMMARY

Saturn launch vehicle SA-9, fourth of the Block II series vehicles and the first operational vehicle, was launched at 9:37 AM EST, February 16 1965. The flight test was the first in a series to launch a Pegasus satellite (Pegasus A) and was a complete success with all missions achieved.

SA-9 was the fourth vehicle launched from complex 37B, Eastern Test Range (ETR), Cape Kennedy, Florida, and represents the third launch of a Saturn/Apollo configuration. As a result of high S-I and S-IV stage performance the actual flight path of SA-9 deviated from nominal.

SA-9

The performance of both S-I and S-IV propulsion systems was satisfactory for the SA-9 flight.

SA-9 was the fourth Saturn vehicle to employ H-1 engines at a thrust level of 836,000N (188,000 lbf) to power the S-I stage. SA-9 also represented the fourth flight of the RL10A-3 engines to power the S-IV stage. The overall performance of the SA-9 guidance and control system was satisfactory.

The vibrations observed on SA-9 were all within the expected levels and compared well with those of SA-7. The configuration of the SA-9 Instrument Unit was changed. As a result the vibration levels during flight were somewhat higher than those experienced in the SA-7 flight. However, this increase was expected.

SA-9 was the first of the Block II vehicles to fly a prototype model of the production Instrument Unit. This IU is environmentally controlled during preparations for flight by the ground support equipment, and uses no in flight environmental control system. The SA-9 electrical systems operated satisfactorily. Overall reliability of the SA-9 measuring system was 98.9 percent. Only 16 of the 1244 measurements on the vehicle at liftoff failed.

Main engine flame attenuation was less severe than on SA-7 and retro rocket effects were greatly improved over previous flights because of higher altitude separation.

The Pegasus A spacecraft performed satisfactorily. At approximately 631.66 seconds, the S-IV stage, Instrument Unit, Apollo shroud, and Pegasus were inserted into orbit with no appreciable pitch, yaw or roll rate. Pegasus wing deployment was successful and all spacecraft systems operated properly. After wing deploy-ment a roll rate started to build up and reached a maximum of 9.8 deg/s at the end of GOX venting. This is believed to have been caused by the venting of GOX impinging on the Pegasus wings.

TEST OBJECTIVES
(All Accomplished)

1. Demonstrate the functional operations of the Pegasus meteoroid technology satellite mechanical. Structural and electronic subsystems
2. Evaluate meteoroid data sampling in near earth orbit.
3. Evaluate closed loop guidance accuracy and demonstrate iterative guidance mode (first flight utilization of iterative guidance scheme)
4. Evaluate S-IV/IU/Service Module adapter (SMA) exterior thermal control coating
5. Demonstrate S-IV stage nonpropulsive venting system
6. Demonstrate boilerplate Command Module (CM)/ SM separation from S-IV/IU/SMA
7. Demonstrate redesigned unpressurized IU and passive thermal control system
8. Demonstrate S-IV propulsion and vehicle control
9. Demonstrate S-I/S-IV separation
10. Demonstrate launch escape system (LES) jettison
11. Demonstrate S-I propulsion and vehicle control
12. Evaluate launch environment

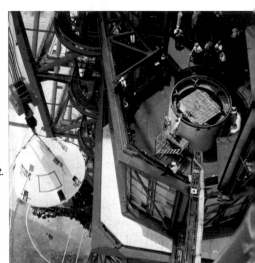

Pegasus B can be seen inside the dummy Apollo service module at right.

SATURN/APOLLO (SA) 08 FLIGHT SUMMARY

Saturn launch vehicle SA-8, fifth of the Block II series vehicles and the second operational vehicle, was launched at 02:35 AM EST, May 25, 1965. The flight test was the second in a series to launch a Pegasus satellite (Pegasus B) and was a complete success with all missions achieved.

SA-8 was the fifth vehicle launched from complex 37B at Cape Kennedy, Florida, and represents the fourth launch of the Saturn/Apollo configuration. This was the first Saturn vehicle launch that required no technical holds.

Two anomalies were detected during the Countdown Operation. The first occurred during countdown when LOX vapor periodically broke the theodolite line of-sight to the ST-124 alignment window in the Instrument Unit (IU). This was the second launch in which LOX vapor temporarily and periodically hindered the Countdown operation. The second anomaly occurred when the GH2 vent disconnect on swing arm 3 failed to separate pneumatically at liftoff. However, separation was successfully achieved when the mechanical release was actuated by the swing

SA-8

arm rotation. The disconnect was accomplished by a hydraulic lanyard during launch. A similar malfunction occurred during the launch of SA-7.

The actual trajectory of SA-8 deviated from nominal primarily because of high S-1 stage performance. The total velocity was 21.8 m/s higher than nominal at OECO and 0.5 m/s lower than nominal at S-IV Cutoff.

The performance of both the S-I and S-IV propulsion systems was satisfactory for the SA-8 flight. The performance of all subsystems was as expected for the flight test. The overall performance of the SA-8 guidance and control system was satisfactory.

Separation of the Apollo shroud occurred 7 seconds, functioning as planned.

The structural flight loads on SA-8 were generally as expected and no POGO effects were apparent. The vibrations observed on SA-8 were generally within the expected levels and compared well with those of SA-9.

The SA-8 electrical systems

operated satisfactorily during the boost and orbital phase of flight and all mission requirements were met.

Overall reliability of the SA-8 measuring System was 99.4 percent. Only 7 of the 1157 measurements on the vehicle at Liftoff failed. The Playback records were free of attenuation effects caused by the retro and ullage rockets. The photo/optical instrumentation consisted of 79 cameras that provided fair quality coverage. Of the 79 cameras, 3 failed, 4 had no

timing, and 4 had unusable timing due to edge fog on the film. The onboard TV system provided excellent coverage of the Pegasus wing deployment. The Pegasus B spacecraft performance was satisfactory. At approximately 634.15 seconds, the SIV stage, Instrument Unit, Apollo shroud and Pegasus were inserted into orbit with no appreciable pitch, yaw, or roll rate. The Pegasus wing deployment and all spacecraft Systems worked properly and all measurements were initially within their predicted limits.

TEST OBJECTIVES
(All accomplishd)
Primary objectives;
1. Evaluate meteoroid data sampling in near earth orbit
2. Demonstrate iterative guidance mode and evaluate system accuracy

Secondary objectives:
1. Demonstrate the functional operation of the Pegasus meteoroid technol-

ogy satellite mechanical, structural, and eleetronic subsystem
2. Evaluate S-IV/IU/Service Module adapter (SMA) exterior thermal control coating
3. Demonstrate boilerplate Command Module (CM)/SM Separation from S-IV/IU/SMA
4. Demonstrate S-IV stage nonpropulsive venting (NPV) system

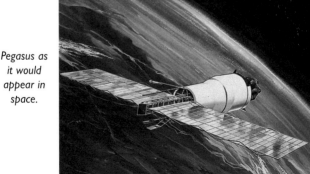

Pegasus as it would appear in space.

SATURN/APOLLO (SA) 10
FLIGHT SUMMARY

Saturn launch vehicle SA-10, sixth of the Block II series vehicles and the third operational vehicle, was launched at 08:00 AM EST, July 30, 1965. This flight test was the tenth and last in a series of Saturn I vehicles to be flight tested. The flight test was the third in a series to launch a Pegasus satellite (Pegasus C) and was a complete success with all mission objectives achieved.

SA-10 was the sixth vehicle launched from complex 37B at Cape Kennedy, Florida, and represents the fifth launch of the Saturn/Apollo configuration. This was the second Saturn vehicle launch that required no technical holds. All operations were normal and the only hold was the 30-minute built-in-hold, used to make launch time coincident with the beginning of the launch window. The major anomaly associated with countdown operations was high surface winds; 8.7 m/s (16.9 knots) were prevalent during the hour preceding launch. The actual trajectory of SA-10 was very close to nominal.

The S-IV stage and payload at orbital insertion (S-IV cutoff plus

SA-10

10 seconds) had a space fixed velocity 0.7 m/s less than nominal, yielding a perigee altitude of 528. 8 km and an apogee altitude of 531.9 km. Estimated orbital lifetime was 720 days, 5 days less than nominal.

The performance of both the S-I and S-IV propulsion systems was satisfactory for the SA-10 flight,

The overall performance of the SA-10 guidance and control systems was satisfactory. Vehicle response to all signals was properly executed including roll maneuver, pitch program, and path guidance (utilizing the iterative guidance scheme) during the S-IV stage flight.

Separation of the Apollo shroud occurred at 812.10 seconds, functioning as planned.

The vibrations observed on SA-10 were generally within the expected levels and compared well with SA-8. The electrical system of SA-10 vehicle operated satisfactorily during boost and orbital phases of flight and all mission requirements were met.

Overall reliability of the SA-10 measuring system was 98.8 percent, considering only those measurements active at liftoff. There were 1018 measurements active at liftoff of which 12 failed during flight. All airborne tape recorders operated satisfactorily. The onboard TV system for SA-10 was cancelled prior to flight. The altimeter system and associated return pulse-shape experiment failed to operate. The MIS-TRAM transponder failed at 63 seconds of flight and provided no usable data. The photo/optical coverage for SA-10 was good.

However, downrange cloud conditions prevented all of the 10.2 m (400 in) and 12.7 m (500 in) focal length cameras from recording usable data.

The Pegasus C spacecraft performance was satisfactory. At approximately 640.252 seconds, the SIV stage, Instrument Unit, Apollo shroud and Pegasus were inserted into orbit with no appreciable pitch, yaw, or roll rate. The Pegasus wing deployment and all spacecraft systems worked properly and all measurements were initially within their predicted limits.

TEST OBJECTIVES
(All Accomplished)
Primary objectives
1. Collection and evaluation of meteoroid data in near earth orbit
a. Determination of meteoroid penetration of satellite panels for three thicknesses of aluminum.
b. Measurement of satellite's radiation environment and panel temperature to evaluate the validity of hit data.
c. Determination of satellite's position and orientation relative to time of hit occurrence.
2. Continued demonstration of launch vehicle iterative guidance mode and evaluation of system accuracy

Secondary objectives
1. Evaluation of the functional operation of the Pegasus meteoroid technology satellite's mechanical structural, and electronics subsystems
2. Evaluation of S-IV/IU/Service Module adapter (SMA) exterior thermal control coating
3. Evaluation of boilerplate Command Module (CM) /SM separation from the S-IV/IU/SMA
4. Evaluation of the S-IV stage nonpropulsive venting system

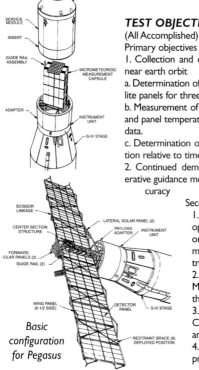

Basic configuration for Pegasus

AS-201 FLIGHT SUMMARY

GENERAL
Spacecraft: CSM-009
Launch Vehicle: SA-201
Launch Complex: 34
Launch Time: 11:12 a.m. EST, February 26, 1966
Launch Azimuth: 105°
Sub-orbital Flight - Maximum Altitude: 266 NM
Mission Duration: 37 minutes 19 seconds
Time of Landing: 11:49 a.m. EST

SPACE VEHICLE AND PRE-LAUNCH DATA
Spacecraft delivered to Cape Kennedy: October 1965
Launch vehicle delivered to Cape Kennedy:
   First stage (S-IB): September 1965
   Second stage (S-IVB): September 1965
   Instrument unit (IU): October 1965
Spacecraft launch weight: 45,900 lb.
Space vehicle weight at liftoff: 1,317,900 lb.

MISSION PRIMARY OBJECTIVES
(All Objectives Accomplished)

1. Demonstrate structural integrity and compatibility of the launch vehicle and confirm launch loads.
2. Test the Separation of:
   a) S-IVB stage, instrument unit (IU), and spacecraft from S-IB stage.
   b) Launch escape system (LES) and boost protective cover from command/Service module and launch vehicle.
   c) CSM from S-IVB stage, IU, and service module-LM adapter (SLA).
   d) Command module (CM) from service module (SM).
3. Obtain flight operation information on the following subsystems:
   a) Launch vehicle: propulsion, guidance and control, and electrical systems.
   b) Spacecraft: CM heat shield (adequacy for entry from low earth orbit); service propulsion system (SPS) (including re-start); environmental control system (ECS) (pressure and temperature control); communications; CM reaction control system (RCS); SM RCS; stabilization control system (SCS); earth landing system (ELS); and electrical power system (EPS).
4. Evaluate performance of the space vehicle emergency detection system (EDS) in an open-loop configuration.
5. Evaluate the CM heat shield at a heating rate of approximately 200 Btu/ft./sec. during entry at approximately 28,000 fps.
6. Demonstrate the mission sup-

port facilities and operations required for launch, mission conduct, and CM recovery.

DETAILED TEST OBJECTIVES
(All Objectives Accomplished)

PRINCIPAL OBJECTIVES
Launch Vehicle:

1. Demonstrate compatibility and structural integrity of the space vehicle (SV) during S-IB stage-powered flight and confirm structural loads and dynamic characteristics.
2. Demonstrate structural integrity and compatibility of S-IVB and space vehicle during powered phase and coast.
3. Demonstrate separation of:
   a) S-IVB from S-IB.
   b) CSM from S-IVB/IU/SLA
4. Demonstrate S-IVB propulsion system including program mixture ratio shift and determine system - performance parameters.
5. Demonstrate S-IB propulsion system and evaluate subsystem performance parameters.
6. Demonstrate launch vehicle guidance system, achieve guidance cutoff & evaluate system accuracy.
7. Demonstrate LV control system during S-IVB-powered phase, S-IVB coast phase, and S-IB-powered phase, and evaluate performance characteristics.
8. Demonstrate LV sequencing system.
9. Evaluate performance of the space vehicle EDS in an open-loop configuration.
10. Demonstrate the mission support facilities required for launch, mission operations and recovery.

Spacecraft:

1. Determine performance of the SCS and determine its adequacy for manned orbital flight.
2. Verify SPS operation for a minimum of 20 seconds after at least 2 minutes in space environment and verify restart capability.
3. Determine performance of the CM RCS and SM RCS to determine their adequacy for manned orbital flight.
4. Determine long duration (approximately 200 seconds) SPS performance including shutdown characteristics.
5. Obtain data on SPS engine firing stability.
6. Determine performance of ECS (pressure and temperature control) and its adequacy for manned orbital flight.
7. Determine performance of the EPS and determine its adequacy for manned orbital flight.
8. Determine performance of the communication System and determine its adequacy for manned orbital flight.
9. Demonstrate compatibility and structural integrity of CSM/Saturn IB.
10. Determine structural loading of SLA when subjected to the Saturn IB launch environment.
11. Demonstrate separation of the S-IVB from the S-IB, the LES and boost protective cover from the CSM, the CSM from the S-IVB/IU/SLA, and the CM from the SM.

12. Determine CM adequacy for manned entry from low earth orbit.

13. Evaluate the CM heat shield ablator at a high heating rate of approximately 200 Btu/ft.2/sec. during entry at 28,000 fps.

14. Demonstrate operation of the parachute recovery subsystem and recovery aids following entry.

SECONDARY OBJECTIVES
Launch Vehicle:

1. Confirm LV-powered flight external environment.

2. Evaluate LV internal environment.

3. Evaluate IU/S-IVB inflight thermal conditioning system,

4. Demonstrate adequacy of S-IVB residual propellant venting System.

UNUSUAL FEATURES OF THE MISSION

1. First flight of the Saturn IB Launch Vehicle with both the S-IB first stage and the S-IVB second stage.

2. First non-orbital flight separation of the launch vehicle and spacecraft in the Saturn IB configuration.

3. First CM recovery.

4. First SPS burn and restart.

5. First flight test of a Block I Apollo Spacecraft.

6. First employment of the Mission Director concept in Apollo.

7. First employment of Mission Control Center - Houston (MCC-H) for Apollo mission control.

Spacecraft differences from "operational" Block I configuration:

* A developmental Block I heat shield was added.

* The guidance and navigation system was omitted.

* An open-loop EDS for the LES was added.

* Couches, space suits, and Crew provisions were omitted.

* Batteries were substituted for fuel cells in the EPS.

* Biomedical instrumentation was omitted in the instrumentation system.

* Certain displays and controls related to astronaut operation were omitted.

* A CM control programmer and attitude reference system was added.

* Additional research and development (R&D) instrumentation was included.

The SA-201 Launch Vehicle was a standard Saturn IB design with the following exceptions:

* R&D instrumentation was included.

* An open-loop EDS was added.

* R&D structure was used in the S-IB stage.

RECOVERY DATA

Recovery Area: Atlantic Ocean

Landing Coordinates: 8°56'S., 10°43'W.

Recovery Ship: USS Boxer

Spacecraft Recovery Time: 2:13 p.m. EST, February 27, 1966

AS-203 FLIGHT SUMMARY

GENERAL
Launch Vehicle: SA-203
Launch Complex: 37
Launch Time: 9:53 a.m. EST, July 5, 1966
Launch Azimuth: 72°
Apogee: 101.8 NM
Perigee: 101.6 NM
Revolutions: 4 (Vehicle broke up during pressure test above design value.) Vehicle recovery was not planned.

SPACE VEHICLE AND PRE-LAUNCH DATA
No spacecraft was carried on this mission. An aerodynamic fairing (nosecone) weighing 3700 lb. was attached to the instrument unit and contained an MSC subcritical cryogenic experiment.

Launch vehicle delivered to Cape Kennedy:
    First stage (S-IB): April 1966
    Second stage (S-IVB): March 1966
    Instrument unit (IU): April 1966
Space vehicle liftoff weight: 1,187,000 lb.
Total weight in orbit: 58,500 lb.

MISSION PRIMARY OBJECTIVES
(All Objectives Accomplished)

1. Evaluate the S-IVB LH2 continuous venting system.

AS-203

2. Evaluate S-IVB engine chilldown and recirculation system.
3. Determine S-IVB tank fluid dynamics.
4. Determine heat transfer into liquid hydrogen (LH2) through tank wall, and obtain data required for propellant thermodynamic model.
5. Evaluate S-IVB and IU checkout in orbit.
6. Demonstrate orbital operation of the launch vehicle attitude control and thermal control systems.
7. Demonstrate the ability of the launch vehicle guidance to insert a payload into orbit.
8. Demonstrate operational structure of the launch vehicle.
9. Demonstrate the mission support facilities and operations required for launch and mission control.

DETAILED TEST OBJECTIVES
(All Objectives Accomplished)

PRINCIPAL OBJECTIVES
Launch Vehicle:
1. Evaluate the J-2 engine LH2 chilldown and re-circulation system, and ullage requirements for simulated engine restart.
2. Determine cryogenic liquid/vapor interface configuration and fluid

dynamics of propellants in near zero-g environment.

3. Demonstrate the S-IVB auxiliary propulsion system operation and evaluate performance parameters.

4. Demonstrate the adequacy of the S-IVB/IU thermal control system.

5. Demonstrate the launch vehicle guidance system operation, achieve guidance cutoff, and determine system accuracy.

6. Demonstrate the structural integrity of the launch vehicle and determine its dynamic characteristics.

SECONDARY OBJECTIVES
Launch Vehicle:

1. Evaluate the launch vehicle-powered flight external environment.

2. Verify the launch vehicle sequencing system operation.

3. Evaluate performance of the EDS in an open-loop configuration..

4. Evaluate separation of S-IVB/IU/nosecone from S-IB.

5. Verify launch vehicle propulsion systems' operation and evaluate system performance parameters.

6. Evaluate the MSC subcritical cryogenic experiment.

UNUSUAL FEATURES OF THE MISSION

1. Simulated S-IVB engine restart in orbit.

2. Use of hydrogen continuous vents to accelerate payload in orbit for settling S-IVB LH2.

3. First orbital flight for S-IVB stage.

4. Insert most weight to date in orbit by the United States (28 tons).

5. Television feedback and behavior of LH2 under orbital conditions.

6. First flight for redesigned, lighter weight S-IB stage.

The SA-203 Launch Vehicle differed from the SA-201 vehicle as follows:

* The S-IB stage weight was decreased by 28,500 lb.
* The S-IB stage had a redesigned propellant container, barrel assembly, outriggers and gaseous oxygen interconnect and vent system.
* The S-IB stage outboard engine skirt was removed.

AS-202 FLIGHT SUMMARY

GENERAL
Spacecraft: CSM-011
Launch Vehicle: SA-202
Launch Complex": 34
Launch Time: 12:15 p.m. EST, August 25, 1966
Launch Azimuth: 105°
Apogee: 617.1 NM No Orbital Insertion Planned.
Mission Duration: 1 hour 33 minutes
Time of Landing: 1:48 p.m. EST

SPACE VEHICLE AND PRE-LAUNCH DATA
Spacecraft delivered to Cape Kennedy: April 1966
Launch vehicle delivered to Cape Kennedy:
  First stage (S-IB): February 1966
  Second stage (S-IVB): January 1966
  Instrument unit (IU): February 1966
Spacecraft launch weight: 56,900 lb.
Space vehicle weight at liftoff: 1,312,300 lb.

MISSION PRIMARY OBJECTIVES
(All Objectives Accomplished)

1. Demonstrate structural integrity and compatibility of the launch vehicle and spacecraft and confirm launch loads.
2. Demonstrate separation of:

a) S-IVB/IU/spacecraft from S-IB.
b) LES and boost protective cover from CSM/launch vehicle.
c) CSM from S-IVB/IU/SLA.
d) CM from SM.
3. Verify operation of the following subsystems:
a) Launch vehicle: propulsion, guidance and control, and electrical systems.
b) Spacecraft: CM heat shield (adequacy for entry from low earth orbit); SPS (including multiple restart); guidance and navigation, environmental control system; communications; CM reaction control system; SM reaction control system; stabilization control system; earth landing system; and electrical power system.
4. Evaluate performance of the space vehicle EDS in closed-loop configuration.
5. Evaluate the heat shield at high heat load during entry at approximately 28,000 fps.
6. Demonstrate the mission support facilities and operations required for launch, mission conduct, and CM recovery.

DETAILED TEST OBJECTIVES
(All Objectives Accomplished)

PRINCIPAL OBJECTIVES
Launch Vehicle:

1. Demonstrate structural integrity and compatibility of the space vehicle during S-IB stage-powered flight and confirm structural loads and dynamic characteristics.

2. Demonstrate structural integrity and compatibility of the space vehicle during S-IVB stage-powered flight and coast.

3. Demonstrate S-IVB propulsion system operation including program mixture ratio shift and evaluate system performance parameters.

4. Demonstrate S-IB propulsion system operation and evaluate system performance parameters.

5. Demonstrate launch vehicle guidance system operation, achieve guidance cutoff, and evaluate system accuracy.

6. Demonstrate launch vehicle control system operation during S-IB-powered phase, S-IVB-powered phase, & S-IVB coast phase; & evaluate performance characteristics.

7. Demonstrate launch vehicle sequencing system operation.

8. Demonstrate the inflight performance of the S-IB and S-IVB secure range command systems.

Spacecraft:

1. Determine performance of guidance and navigation subsystem and its adequacy for a manned orbital mission.

2. Evaluate guidance and navigation performance during boost and closed-loop entry.

3. Determine performance of the SCS and determine its adequacy for manned orbital flight.

4. Demonstrate multiple SPS restart (at least three burns of at least three-second duration at ten second intervals.)

5. Evaluate performance of the CM RCS and the SM RCS to determine their adequacy for manned orbital flight.

6. Verify SPS standpipe fix (minimum of 198 seconds of SPS burn required.)

7. Determine long duration (approximately 200 seconds) SPS performance including shutdown characteristics.

8. Obtain data on SPS engine firing stability.

9. Determine performance of ECS and its adequacy for manned orbital flight.

10. Determine performance of the EPS and determine its adequacy for manned orbital flight.

11. Determine performance of the communication system and determine its adequacy for manned orbital flight.

12. Verify S-band communications operations for turn-around ranging mode and downlink modes.

13. Demonstrate compatibility and structural integrity of CSM/Saturn IB.

14. Determine CM adequacy for manned entry from low earth orbit.

15. Verify astrosextant thermal protection subsystem.

16. Demonstrate operation of the parachute recovery subsystem and recovery aids following reentry.

SECONDARY OBJECTIVES
Launch Vehicle:
1. Confirm launch vehicle-powered flight external environment.
2. Evaluate IU/S-IVB inflight thermal conditioning system.
3. Verify adequacy of S-IVB residual propellant venting.
4. Evaluate the S-IVB common bulkhead reversal test.

UNUSUAL FEATURES OF THE MISSION
1. First use of fuel cells in the Service module on an Apollo/Saturn flight.
2. First flight of the emergency detection system in closed-loop configuration.
3. First recovery of Apollo spacecraft in Pacific area.
4. First test of unified S-band communications.
5. Repeat of the second stage (S-IVB) common bulkhead pressure test.
6. "Black out" communication test.
7. First flight of Apollo guidance and navigation system.

Spacecraft 011 differences from the "operational" Block I configuration:

* A developmental Block I heat shield was added.

* Couches, space suits, and crew provisions were omitted.
* A tie-bar to replace a lunar module was added.
* The S-band in the communication system was omitted:
* Biomedical instrumentation was omitted in the instrumentation system.
* Certain displays and controls related to astronaut operation were omitted.
* A CM control programmer and attitude reference system was added.
* Additional R&D instrumentation was included.

The SA-202 Launch Vehicle differed from the standard Saturn IB design as follows:

* R&D instrumentation was included.
* R&D structure for S-IB stage was included.
* TV camera was included in the IU to view CSM Separation.

RECOVERY DATA
Recovery Area: Pacific Ocean
Landing Coordinates: 16°7'N., 168°54'E.
Recovery Ship: USS Hornet
Spacecraft Recovery Time: 10:10 p.m. EST, August 25, 1966

APOLLO I (AS-204) SUMMARY

The following information on Apollo 1 outlines what was scheduled to happen during the flight. Apollo 1 never left the pad because a catastrophic fire consumed the Command Module during a routine ground test. The flight crew did not survive the fire.

GENERAL

Spacecraft:
CM-012, SM-012
Launch Vehicle: SA-204
Launch Complex: 34
Flight Crew:
  Commander:
  Virgil I. Grissom
  Senior Pilot:
  Edward H. White, II
  Pilot: Roger B. Chaffee
Launch Time (Scheduled): Between 10 a.m. & 3.30 pm EST February 21, 1967
Launch Azimuth (predicted): 72°
Apogee (predicted): 130 NM
Perigee (predicted): 85 NM
Mission Duration (scheduled): Up to 14 days

SPACE VEHICLE AND PRE-LAUNCH DATA

Spacecraft delivered to KSC:
August 26 1966
Launch vehicle delivered to Cape Kennedy:
  First stage (S-IB): August 15 1966
  Second stage (S-IVB): August 7 1966
  Instrument Unit (IU): August 1966
Spacecraft weight: 38,600 lb.
Launch vehicle weight: 1,300,000 lb. (approx)

MISSION PRIMARY OBJECTIVES

1. Verify spacecraft/crew operations.
2. Determine CSM subsystem performance in earth orbital environment.
3. Evaluate S-IVB and IU checkout in orbit.
4. Demonstrate the adequacy of the launch vehicle attitude control system for orbital operation.
5. Demonstrate crew/CSM/launch vehicle/mission support facilities performance during an earth orbital mission.

DETAILED OBJECTIVES & EXPERIMENTS

No operational tests were to be performed on this mission. The following priority sequence for in-flight experiments was to be used as a guide in both pre flight and real-time decisions:

S-051 Daytime Sodium Cloud
M-048 Cardiovascular Reflex Conditioning
M-009 Human Otolith Function
M-003 In-Flight Exerciser
T-003 In-Flight Nephelometer
M-004 In-Flight Phonocardiogram

S-028 Dim Light Photography
S-005 Synoptic Terrain Photography
S-006 Synoptic Weather
Photography

Pre- and post flight experiments M-011 (Cytogenetic Blood Studies) and M-006 (Bone Demineralization) will also be performed.

UNUSUAL FEATURES OF THE MISSION
1. First manned Apollo flight.
2. First flight of the Apollo space suits.
3. First flight with full crew support equipment.

REMARKS
All Mission objectives were unsuccessful.

On January 27, 1967, tragedy struck the Apollo program when a flash fire occurred in command module 012 during a launch pad test of the Apollo/Saturn space vehicle being prepared for the first piloted flight, the AS-204 mission. Three astronauts, Lt. Col. Virgil I. Grissom, a veteran of Mercury and Gemini missions; Lt. Col. Edward H. White, the astronaut who had performed the first United States extravehicular activity during the Gemini program; and Roger B. Chaffee, an astronaut preparing for his first space flight, died in this tragic accident. A seven-member board conducted a comprehensive investigation to pinpoint the cause of the fire. The final report presented the results of the investigation and made specific rec-ommendations that led to major design and engineering modifications, and revisions to test planning, test discipline, manufacturing processes and procedures, and quality control. With these changes, the overall safety of the command and service module and the lunar module was increased substantially. The AS-204 mission was redesignated Apollo 1 in honor of the crew.

SUMMARY OF REVIEW BOARD FINDINGS

* There was a momentary power failure at 23:30:55 GMT.
* Evidence of several arcs was found in the post-fire investigation.
* No single ignition source of the fire was conclusively identified.
* The Command Module contained many types and classes of combustible material in areas contiguous to possible ignition sources.
* The test was conducted with a 16.7 pounds per square inch absolute, 100-percent oxygen atmosphere.

The most probable initiator was an electrical arc in the sector between -Y and +Z spacecraft axes. The exact location best fitting the total available information is near the floor in the lower forward section of the left-hand equipment bay where Environmental Control System (ECS) instrumentation power wiring leads into the area between the Environmental Control Unit (ECU) and the oxygen panel. No evidence was discovered that suggested sabotage.

* The test conditions were extremely hazardous.
* The overall communication system was unsatisfactory.
* Adequate safety precautions were neither established nor observed for this test.

Deficiencies existed in Command Module design, workmanship and quality control, such as:

* Components of the Environmental Control System installed in Command Module 012 had a history of many removals and of technical difficulties including regulator failures, line failures and Environmental Control Unit failures. The design and installation features of the Environmental Control Unit makes removal or repair difficult.
* Coolant leakage at solder joints has been a chronic problem.
* The coolant is both corrosive and combustible.
* Deficiencies in design, manufacture, installation, rework and quality control existed in the electrical wiring.
* No vibration test was made of a complete flight-configured space-craft.
* Spacecraft design and operating procedures currently require the disconnecting of electrical connections while powered.
* No design features for fire protection were incorporated.

These deficiencies created an unnecessarily hazardous condition and their continuation would imperil any future Apollo operations.

* The rapid spread of fire caused an increase in pressure and temperature which resulted in rupture of the Command Module and creation of a toxic atmosphere. Death of the crew was from asphyxia due to inhalation of toxic gases due to fire. A contributory cause of death was thermal burns.
* Non-uniform distribution of carboxyhemoglobin was found by autopsy.

Due to internal pressure, the Command Module inner hatch could not be opened prior to rupture of the Command Module. The crew was never capable of effecting emergency egress because of the pressurization before rupture and their loss of consciousness soon after rupture.

APOLLO 4 (AS-501)
FLIGHT SUMMARY

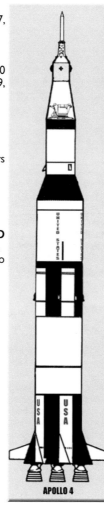

APOLLO 4

GENERAL
Spacecraft: CSM-107, LTA-10R
Launch Vehicle: SA-501
Launch Complex: 39A
Launch Time: 7:00:00 a.m. EST, November 9, 1967
Launch Azimuth: 72°
Apogee: 9767 NM
Perigee: 100 NM
Revolutions: 3
Mission Duration: 8 hrs 37 mins 08 secs
Time of Landing: 3:38:09 p.m. EST, November 9, 1967

SPACE VEHICLE AND PRE-LAUNCH DATA
Spacecraft delivered to KSC:
  Command/service module: Dec. 1966
  Lunar module test article: Sept. 1966
Launch vehicle delivered to KSC:
  First stage (S-IC): September 1966
  Second stage (S-II): January 1967
  Third stage (S-IVB): August 1966
  Instrument unit (IU): August 1966
  Spacecraft weight at liftoff: 93,700 lb.
  Space vehicle weight at liftoff: 6,121,466 lb.

MISSION PRIMARY OBJECTIVES
(All Objectives Accomplished)
1. Demonstrate the structural and thermal integrity and compatibility of the launch vehicle and spacecraft. Confirm launch loads and dynamic characteristics.
2. Demonstrate separation of:
a) S-II from S-IC (dual plane).
b) S-IVB from S-II.
3. Verify operation of the following subsystems:
a) Launch vehicle: propulsion (including S-IVB restart), guidance and control, and electrical system.
b) Spacecraft: CM heat shield, (adequacy of Block II design for entry at lunar return conditions); and selected subsystems.
4. Evaluate performance of the space vehicle EDS in an open-loop configuration.
5. Demonstrate mission support facilities and operations required for launch, mission conduct, and CM recovery.

DETAILED TEST OBJECTIVES
PRINCIPAL OBJECTIVES
Launch Vehicle:

1. Demonstrate the S-IVB stage restart capability.
2. Demonstrate the adequacy of the S-IVB continuous vent system while in earth orbit.
3. Demonstrate the capability of the S-IVB auxiliary propulsion system during S-IVB-powered flight and orbital coast periods to maintain attitude control and perform required maneuvers.
4. Demonstrate the S-IVB stage propulsion system, including the propellant management systems, and determine inflight performance parameters.
5. Demonstrate the S-II stage propulsion system, including programmed mixture ratio shift and the propellant management system, and determine inflight performance parameters.
6. Demonstrate the S-IC stage propulsion system, and determine inflight system performance parameters.
7. Demonstrate S-IC/S-II dual plane separation.
8. Demonstrate S-II/S-IVB separation.
9. Demonstrate the mission support capability required for launch and mission operations to high post injection altitudes.
10. Demonstrate structural and thermal integrity of the launch vehicle throughout powered and coasting flight, and determine inflight structural loads and dynamic characteristics.
11. Determine inflight launch vehicle internal environment.
12. Demonstrate the launch vehicle guidance and control system during S-IC, S-II, and S-IVB powered flight; achieve guidance cutoff; and evaluate System accuracy.
13. Demonstrate launch vehicle sequencing system.
14. Evaluate the performance of the emergency detection system in an open-loop configuration.
15. Demonstrate compatibility of the launch vehicle and spacecraft.
16. Verify prelaunch and launch support equipment compatibility with launch vehicle and spacecraft Systems.

Spacecraft:

1. Verify operation of the guidance and navigation system after subjection to the Saturn V boost environment.
2. Verify operation of the guidance and navigation system in the space environment after S-IVB separation.
3. Verify operation of the guidance and navigation/SCS during entry and recovery.
4. Gather data on the effects of a long duration SPS burn on spacecraft stability.
5. Demonstrate SPS no-ullage start.
6. Determine performance of the SPS during a long duration burn.
7. Verify operation of the CM RCS during entry and throughout the mission.
8. Verify operation of the heat rejection system throughout the mission.
9. Verify operation of the EPS after being subjected to the Saturn V launch environment.
10. Verify operation of the primary guidance system (PGS) after being subjected to the Saturn V launch environment.
11. Verify operation of the EPS in the space environment after S-IVB separation.
12. Verify operation of the PGS in the space environment after S-IVB separation.
13. Verify operation of the EPS during entry and recovery.
14. Demonstrate the performance of

CSM/MSFN S-band communications.

15. Demonstrate satisfactory operation of CSM communication subsystem using the Block II-type VHF omnidirectional antennas.

16. Obtain data via CSM-ARIA communications.

17. Demonstrate CSM/ SLA/ LTA/ Saturn V structural compatibility and determine spacecraft loads in a Saturn V launch environment.

18. Determine the dynamic and thermal responses of the SLA/CSM structure in the Saturn V launch environment.

19. Evaluate the thermal and structural performance of the Block II thermal protection system, including effects of cold soak and maximum thermal gradient when subjected to the combination of a high heat load and a high heating rate representative of lunar return entry.

20. Verify the performance of the SM RCS thermal control subsystem and engine thermal response in the Jeep space environment.

21. Verify the thermal design adequacy of the CM RCS thrusters and extensions during simulated lunar return entry.

22. Evaluate the thermal performance of a gap and seal configuration simulating the unified crew hatch design, for heating conditions anticipated during lunar return entry.

23. Perform flight test of low density ablator panels.

24. Determine the force inputs to the simulated LM from the SLA at the spacecraft attachment structure in a Saturn V launch environment.

25. Obtain data on the acoustic and thermal environment of the SLA/simulated LM interface during a Saturn V launch.

26. Obtain data on the temperature of the simulated LM skin during launch.

27. Determine vibration response of LM descent stage engine and propellant tanks in a Saturn V launch environment.

28. Evaluate the performance of the spacecraft emergency detection system in the open-loop configuration.

29. Verify operation of the ELS during entry and recovery.

30. Measure the integrated skin and depth radiation dose within the CM up to an altitude of at least 2000 NM.

31. Determine the radiation shielding effectiveness of the command module.

32. Determine and display, in real time, Van Allen Belt radiation dose data at the Mission Control Center.

33. Obtain motion pictures for study of entry horizon reference, boost protective cover jettison, and orbit insertion; obtain photographs for earth landmark identification.

SECONDARY OBJECTIVES
Launch Vehicle:

1. Determine launch vehicle-powered flight external environment.

2. Determine attenuation effects of exhaust flames on RF radiating and receiving systems during main engine, retro and ullage motor firings.

UNUSUAL FEATURES OF THE MISSION

1. First launch from LC-39.
2. First flight of Saturn V vehicle.
3. First flight of S-IC stage.
4. First flight of S-II stage.
5. First flight of a lunar module test article (LTA).
6. First orbital restart of S-IVB.
7. First SPS no-ullage start.
8. First simulated Block II heat shield.
9. First lunar return velocity CM reentry.

10. First command and communication system flight test.

11. First use of Apollo Range Instrumentation Aircraft (ARIA).

12. First use of Apollo-configured ships.

Spacecraft differences from previous Block I flights:

* The EDS system operated in open-loop configuration.

* Block II thickness, thermal coating, and manufacturing technique for the CM

heat shield ablator was used.

* A simulated Block II umbilical was added on CM in addition to active Block I umbilical.

* An Apollo Mission Control Programmer with special interface equipment for operation with CSM subsystems was installed in CM in place of crew couches.

* All S-band transmissions and receptions were performed by four S-band omnidirectional antennas modified to reflect Block II configuration.

* Flight qualification tape recorder and associated equipment for R&D measurements were added.

* Couches, crew restraints, crew provisions, instrument panel (partial), SCS (partial), and ECS (partial) were deleted from Block I configuration.

* CM hatch window was replaced with instrumentation test panel containing simulations of flexible thermal seals designed for the developmental quick operating hatches.

* Selected ECS water-glycol joints were armor-plated to evaluate their behavior during a space vehicle launch.

* The CM cabin was filled with gaseous nitrogen (GN2). at liftoff to preclude the possibility of cabin fire.

* CM underwent extensive inspection and rework of its wiring to provide better wiring protection. The lunar module test article (LTA-10R) was a "boilerplate" LM test article instrumented to measure vibration, acoustics, and structural integrity at 36 points in the spacecraft-LM adapter (SLA). Data was telemetered to the ground stations during the first 12 minutes of flight. The LTA-10R used a flight-type descent stage without landing gear. Its propellant tanks were filled with water/ glycol and freon to simulate fuel and oxidizer, respectively. The ascent stage was a ballasted aluminum structure containing no flight systems.

Launch vehicle differences from lunar mission configuration:

* The second stage (S-II) did not have the light weight structure to be used for the lunar mission.

* The F-1 and J-2 engines were not uprated versions.

* The EDS system was in open-loop configuration.

* The O2H2 burner, used as helium heater on S-IVB, was not installed.

RECOVERY DATA

Recovery Area: Pacific Ocean

Landing Coordinates: 30°N, 172°W.

Recovery Ship: USS Bennington

Spacecraft Recovery Time: 5:52 p.m. EST, November 9, 1967

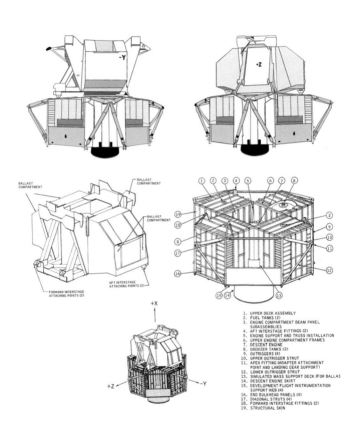

1. UPPER DECK ASSEMBLY
2. FUEL TANKS (2)
3. ENGINE COMPARTMENT BEAM PANEL
   SUBASSEMBLIES
4. AFT INTERSTAGE FITTINGS (2)
5. ENGINE SUPPORT AND TRUSS INSTALLATION
6. UPPER ENGINE COMPARTMENT FRAMES
7. DESCENT ENGINE
8. OXIDIZER TANKS (2)
9. OUTRIGGERS (4)
10. UPPER OUTRIGGER STRUT
11. APEX FITTING (ADAPTER ATTACHMENT
    POINT AND LANDING GEAR SUPPORT)
12. LOWER OUTRIGGER STRUT
13. SIMULATED MASS SUPPORT DECK (FOR BALLAS
14. DESCENT ENGINE SKIRT
15. DEVELOPMENT FLIGHT INSTRUMENTATION
    SUPPORT WEB (4)
16. END BULKHEAD PANELS (4)
17. DIAGONAL STRUTS (4)
18. FORWARD INTERSTAGE FITTINGS (2)
19. STRUCTURAL SKIN

*Basic overview of the Lunar Module Test Article
(LTA) used for Apollo 4 and Apollo 6*

APOLLO 5 (SA-204/LM-1) FLIGHT SUMMARY

GENERAL
Lunar Module: LM-1
Launch Vehicle: SA-204
Launch Complex: 37B
Launch Time: 5:48:08 p.m.
EST, January 22, 1968
Launch Azimuth: 72°
Apogee: 519 NM
Perigee: 88 NM
Mission Duration: 7 hours 50 minutes

SPACE VEHICLE AND PRE-LAUNCH DATA
Spacecraft delivered to KSC:
  Lunar module. (LM-1) :
  June 1967
  Spacecraft-LM Adapter
  (SLA) : October 1966
Launch vehicle delivered to KSC:
  First stage (S-IB):
  July 1966
  Second stage (S-IVB):
  August 1966
  Instrument unit (IU):
  August 1966
Spacecraft launch weight: 31,700 lb.
Space vehicle weight at liftoff: 1,285,400 lb.

MISSION PRIMARY OBJECTIVES
(All Accomplished)

1. Verify operation of the following LM subsystems: Ascent propulsion system and descent propulsion system (including restart), and structure.
2. Evaluate LM staging.
3. Evaluate the S-IVB/IU orbital performance.

DETAILED TEST OBJECTIVES. PRINCIPAL AND MANDATORY OBJECTIVES

Spacecraft:
1. Verify descent engine gimballing response to control signals. (Accomplished)
2. Demonstrate PGNCS thrust vector control and attitude control capability and evaluate the performance of the DAP and IMU in a flight environment. (Partially Accomplished)
3. Determine DPS and APS Start, restart and shutdown characteristics in a space environment. (Accomplished).
4. Verify DPS thrust response to throttling control signals. (Partially Accomplished)
5. Determine that no adverse interactions exist between propellant slosh, vehicle stability, engine vibration and APS/DPS performance. (Partially Accomplished)
6. Determine that no vehicle degradation exists which would affect Crew safety during APS burn to depletion. (Partially Accomplished.)
7. Verify the operation of the DPS propellant feed and pressurization sections. (Partially Accomplished)

APOLLO 5

SECONDARY OBJECTIVES
Launch Vehicle:
1. Evaluate the launch vehicle attitude control system operation and maneuvering capability. (Accomplished)
2. Verify the S-IVB LH2 and LOX tank pressure rise rates. (Accomplished)
3. Demonstrate nosecone separation from the S-IVB/IU/SLA. (Accomplished)
4. Evaluate the operational adequacy of the launch vehicle systems, including guidance and control, electrical, mechanical, and instrumentation. (Accomplished)

Spacecraft:
(All Accomplished)
1. Verify satisfactory operation of portions of the LM ECS equipment.
2. Evaluate the performance of the spacecraft jettison controller (SJC) and pyrotechnical devices in the execution of nose cap Separations, SLA panel deployment and LM/SLA Separation functions.
3. Verify performance of portions of the LM S-band communications subsystem and its compatibility with MSFN.
4. Evaluate the performance of the instrumentation Subsystem during boost and LM propulsion subsystem operations.
5. Demonstrate the operation of the explosive devices.

UNUSUAL FEATURES OF THE MISSION
1. First flight to verify operation of LM subsystems.
2. First firing in space of LM descent engine.
3. First firing in space of LM ascent engine.
4. First test of LM fire-in-the-hole (FITH) staging capability.

Lunar module differences from future LM's:
* An LM mission programmer (LMP) was added to perform control functions normally accomplished by the flight Crew. The LMP received commands from the LM guidance computer (LGC), ground controller or its component program reader assembly (PRA). The PRA contained 64 taped contingency programs to be used in event of LGC failure. The digital command assembly (DCA) provided an uplink capability for routing of ground signals to the LGC for the PRA. The program coupler assembly (PCA) provided coupling of the LGC and PRA commands to the subsystems.
* Developmental flight instrumentation (DFI) was within the LM-1 to supply operational data for flight conditioning electronics, modulationpackages, VHF transmitters, and C-band beacons.
* The lunar mission erectable S-band antenna was not used.
* The mission did not employ a tape recorder for either systems, data, or voice.
* Cable and reel assemblies were used to verify and evaluate (postflight) the ascent/descent stage separation.
* No EVA equipment was used or tested.
* LM guidance was active at liftoff. Normally, this is crew-initiated in a later flight phase. Because this equipment was active at liftoff, the

cooling system was also active.

* This mission employed a space-craft-LM adapter (SLA) umbilical. The LM and SLA were closed out several hours before launch.

* Because LM guidance was activated at liftoff, a guidance reference release signal (GRRS) was transmitted from MCC at approximately T-3 minutes (before automatic count-down sequencing).

* Landing gear was not attached.

* No crew provisions were included.

* Partial deletions were made in the environmental control system (ECS).

* The rendezvous radar was inoperative.

* The two LM cabin windows and the overhead docking window were replaced by aluminum panels.

The SA-204 Launch Vehicle was similar to the previous Saturn IB vehicles.

RECOVERY DATA

No recovery was planned.

REMARKS

An unscheduled hold of 3 hours 48 minutes occurred during the countdown at T-2 hours 30 minutes. The hold was caused by two problems: a failure in the freon supply in the ECS ground support equipment, and a power supply failure in the DDAS. The flight of the SA-204 Launch Vehicle was according to plan. The LM-1 spacecraft also performed according to plan until the time of the first descent propulsion

engine burn. The engine started as planned but was shut down after slightly more than four seconds by the LM guidance subsystem when the velocity did not build up at the predicted rate. The problem was analyzed and was determined to involve guidance Software only, and the decision was made to go to an alternate mission plan that provided for accomplishing the minimum requirements necessary to meet the primary objectives of the mission. The major difference between the planned and alternate missions was the deletion of a long (12-minute) DPS burn and the Substitution of program reader assembly (PPA) control for primary guidance control during the propulsion burns. During all burns conducted under PRA control, there was no attitude control; only rate damping was provided. The alternate plan was successfully executed by the flight operations team. Although not all spacecraft detailed test objectives were fully accomplished, sufficient data were obtained to proceed with the mission schedule.

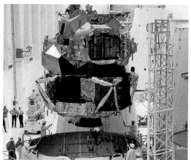

The Lunar Module for Apollo 5. Notice the lack of legs.

APOLLO 6 (AS-502), FLIGHT SUMMARY

GENERAL

Spacecraft: CM-020, SM-014, LTA-2R
Launch Vehicle: SA-502
Launch Complex: 39A
Launch Time: 7:00:00 a.m. EST, April 4, 1968
Launch Azimuth: 72°
Apogee: 12,010 NM
Revolutions: 3
Mission Duration: 9 hours 57 minutes
Time of Landing: 4:57 p.m. EST, April 4, 1968

SPACE VEHICLE AND PRE-LAUNCH DATA

Spacecraft delivered to KSC:
Command/service module (CSM): Nov 1967
Lunar module test article (LTA): February 1967
Launch vehicle delivered to KSC:
First stage (S-IC): March 1967
Second stage (S-II): May 1967
Third stage (S-IVB): February 1967
Instrument unit (IU): March 1967
Spacecraft weight at liftoff: 93,885 lb.
Space vehicle weight at liftoff: 6,108,128 lb.

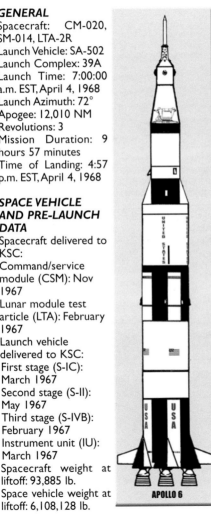

APOLLO 6

MISSION PRIMARY OBJECTIVES

1. Demonstrate the structural and thermal integrity and compatibility of the launch vehicle and spacecraft. Confirm launch loads and dynamic characteristics. (Partially Accomplished)
2. Demonstrate separation of:
a. S-II from S-IC (dual plane). (Accomplished)
b. S-IVB from S-II. (Accomplished)
3. Verify operation of the following launch vehicle subsystems: propulsion (including S-IVB restart), guidance and control (optimum injection), and electrical system. (Partially Accomplished)
4. Evaluate performance of the space vehicle EDS in a closed-loop configuration. (Accomplished)
5. Demonstrate mission support facilities and operations required for launch, mission conduct, and CM recovery. (Accomplished)

DETAILED TEST OBJECTIVES PRINCIPAL AND MANDATORY OBJECTIVES

Launch Vehicle:

1. Demonstrate structural and thermal integrity of launch vehicle throughout powered and coasting flight, and determine inflight structural loads and dynamic characteristics. (Partially Accomplished)

2. Determine inflight launch vehicle internal environment. (Accomplished)

3. Verify pre-launch and launch support equipment compatibility with launch vehicle and spacecraft systems. (Accomplished)

4. Demonstrate the S-IC stage propulsion system and determine inflight system performance parameters. (Accomplished)

5. Demonstrate the S-II stage propulsion system, including programmed mixture ratio shift and the propellant management systems, and determine in-flight system performance parameters. (Partially Accomplished)

6. Demonstrate the launch vehicle guidance and control system during S-IC, S-II, and S-IVB-powered flight. Achieve guidance cutoff and evaluate system accuracy. (Partially Accomplished.)

7. Demonstrate S-IC/S-II dual plane separation. (Accomplished)

8. Demonstrate S-II/S-IVB separation. (Accomplished)

9. Demonstrate launch vehicle sequencing system. (Accomplished)

10. Demonstrate compatibility of the launch vehicle and spacecraft. (Partially Accomplished)

11. Evaluate performance of the emergency detection system (EDS) in a closed-loop configuration. (Accomplished)

12. Demonstrate the capability of the S-IVB auxiliary propulsion system during S-IVB-powered flight and orbital coast periods to maintain attitude control and perform required maneuvers. (Accomplished)

13. Demonstrate the adequacy of the S-IVB continuous vent system while in earth orbit. (Accomplished)

14. Demonstrate the S-IVB stage restart capability. (Not Accomplished)

15. Demonstrate the mission support capability required for launch and mission operations to high post-injection altitudes. (Partially Accomplished)

16. Demonstrate the S-IVB stage propulsion system including the propellant management system, and determine inflight system performance parameters. (Partially Accomplished)

Spacecraft:

1. Evaluate the thermal and structural performance of the Block II thermal protection system, including effects of cold soak and maximum thermal gradient when subjected to the combination of a high heat load and a high heating rate representative of lunar return entry. (Accomplished)

2. Evaluate the thermal performance of a gap and seal configuration simulating the unified crew hatch design for heating conditions anticipated during lunar return entry. (Accomplished)

3. Demonstrate CSM/SLA/LTA/Saturn V structural compatibility and determine spacecraft loads in a Saturn V launch environment. (Partially Accomplished)

4. Determine the dynamic and thermal responses of the SLA/CSM structure in the Saturn V launch environment. (Accomplished)

5. Determine the force inputs to the simulated LM from the SLA at the spacecraft attachment structure in a

Saturn V launch environment. (Accomplished)
6. Evaluate the performance of the spacecraft emergency detection subsystem (EDS) in the open-loop configuration. (Accomplished)
7. Obtain data on the acoustic and thermal environment of the SLA/simulated LM interface during a Saturn V launch. (Accomplished)
8. Determine vibration response of LM descent stage engine and propellant tanks in a Saturn V launch environment. (Accomplished)
9. Demonstrate an SPS no-ullage start. (Accomplished)
10. Verify the performance of the SM RCS thermal control subsystem and engine thermal response in the Jeep space environment. (Accomplished)
11. Verify the thermal design adequacy of the CM RCS thrusters and extensions during simulated lunar return entry. (Accomplished)
12. Verify operation of the heat rejection system throughout the mission. (Accomplished)
13. Measure the integrated skin and depth radiation dose within the command module up to an altitude of at least 2000 nautical miles. (Accomplished)
14. Determine performance of the SPS during a long duration burn. (Accomplished)
15. Demonstrate the performance of CSM/MSFN S-band communications. (Partially Accomplished)

SECONDARY OBJECTIVES

Launch Vehicle:
1. Determine launch vehicle-powered flight external environment. (Accomplished)

2. Determine attenuation effects of exhaust flames on RF radiating and receiving systems during main engine, retro, and ullage motor firings. (Accomplished)

Spacecraft: (All accomplished)
1. Determine and display, in real time, Van Allen belt radiation dose rate and integrated dose data at the Mission Control Center, Houston, Texas.
2. Verify operation of the PGS in the space environment after S-IVB separation.
3. Demonstrate satisfactory operation of CSM communication subsystem using the Block II-type VHF omnidirectional antennas.
4. Verify operation of the G&N/SCS during entry and recovery.
5. Verify operation of PGS after being subjected to the Saturn V launch environment. (Accomplished)
6. Gather data on the effects of a long duration SPS burn on spacecraft stability.
7. Verify operation of the CM RCS during entry and recovery. (Accomplished)
8. Verify operation of the ELS during entry and recovery.
9. Verify operation of the electrical Power system in the space environment after S-IVB separation.
10. Verify operation of the G&N system after subjection to the Saturn V boost environment.
11. Verify operation of the electrical Power system during entry and recovery.
12. Verify operation of the G&N in the space environment after S-IVB separation.
13. Verify operation of the EPS after being subjected to the Saturn V launch environment.
14. Determine the radiation shielding

effectiveness of the CM.

15. Obtain data on the temperature of the simulated LM skin during launch.

16. Obtain data via CSM-ARIA communications.

UNUSUAL FEATURES OF THE MISSION

1. First flight of the emergency detection system (EDS) in a closed-loop configuration.

2. First mission where flight controllers were not deployed to remote sites.

3. First flight of CM unified hatch.

Spacecraft changes from Apollo 4:

* The emergency detection system (EDS) was flown in its normal or "closed-loop" configuration with automatic abort capability.

* The command module contained the new unified, quick operating crew hatch.

* Entry batteries A and B in the CM each had a redundant battery added in parallel in order to eliminate a single-point failure mode.

* On the CM, the thermal coating used on Apollo 4 was replaced with a high emissivity paint in order to simulate the structural temperatures that will be encountered on a lunar mission.

* The micrometeoroid protection windows were removed from the CM.

* Five of the seven operational Block II EVA handrails were installed on the CM. Only two handrails were installed on Apollo 4.

* Five test samples of low-density ablative heat shield materials were flown to test materials which may result in weight savings on future Block II CM's.

Three samples were mounted in place of the left side window and two samples were mounted in the simulated Block II umbilical cavity.

* A 16mm movie camera was added to the CM, positioned to sight out the left rendezvous window to record LES jettison, and to determine visibility of the horizon, window degradation, and plasma brilliance during entry. The 70mm sequence camera used on Apollo 4 was relocated to sight out the Crew hatch window for earth landmark photography.

* Dosimeters were added in the CM to provide evaluation of the operational system for determining crew radiation dose rate and displaying this data in real time at the Mission Control Center.

* A microphone was installed to determine the noise level in the CM during Saturn V launch with the unified crew hatch installed. The CM postlanding vent valve was replaced with the Block II valve.

* The ECS 2.40 controller was replaced with an improved unit having reduced EMI susceptibility, improved potting, and circuitry changes for increased reliability.

* The instrumentation signal mechanical commutators used on Apollo 4 were replaced with solid state commutators having a higher reliability.

* Electrical bonding straps were installed across the CM/SM and LTA/SLA Interfaces to provide electrical bonding without special preparation of mating structural surfaces.

* The SM aft bulkhead was strengthened to have a safety factor of 1.5 at 4.58 g.

* The SPS propellant tank skirt in the SM was strengthened.

* The titanium lines connected to the cryogenic hydrogen tanks in the SM were replaced with stainless steel line and bimetallic adapters.
* The Block I SM RCS engines in Quad B were replaced with Block II engines.
* The SM had the standard Block I white paint whereas the Apollo 4 SM was painted with the Block II aluminized paint.
* The LTA had the landing gear installed permanently in the retracted position.

Launch vehicle differences from the lunar configuration:

* The second stage (S-II) did not have the lightweight structure which will be used with the lunar configuration.
* Neither the F-1 nor the J-2 engine was the uprated version.
* The O2H2 burner used as a helium heater on the S-IVB was not installed.
* R&D instrumentation was installed on all stages.
* The S-IC had two TV cameras looking at the F-1 engines.
* Recoverable cameras were mounted on the S-IC and S-II stages.

RECOVERY DATA

Recovery Area: Pacific Ocean
Landing Coordinates: 27°40'N, 157°59'W.
Recovery Ship: USS Okinawa
Spacecraft Recovery Time: 5:55 p.m. EST, April 4, 1968

REMARKS

During the first stage burn a propulsion structural longitudinal coupling (POGO effect) was noted. At approximately 134 seconds GET all LTA instrumentation showed a sudden unexpected change in dynamic characteristics and airborne lightweight optical tracking system (ALOTS) photos showed debris coming from the SLA area. The S-IC / S-II dual plane separation occurred normally. Approximately 260 seconds after S-II ignition, engines #2 and #3 cut off prematurely. The remaining engines maintained vehicle control through the subsequent portion of the S-II burn. This malfunction caused the S-II stage to burn approximately 58 seconds longer than the nominal time. The S-IVB/S-II separation therefore occurred approximately 59 seconds later than nominal. The first S-IVB burn was approximately 29 seconds longer than nominal due to the S-II malfunction and the subsequent automatic attempt to achieve the proper orbit conditions. Despite the unplanned usage of propellants during the first S-IVB burn, the vehicle loading had sufficient margin that the planned full duration translunar injection burn was still possible. The S-IVB restart sequence was initiated at the end of the second revolution, but the stage failed to complete the ignition sequence. Due to the failure of the S-IVB to reignite, an alternate mission was selected. This mission consisted of firing the service propulsion system (SPS) to attain the planned apogee of approximately 12,000 NM. To achieve this altitude a burn duration of 445 seconds was required, leaving residuals sufficient for a second burn of only 23 seconds. Because of this low propellant quantity, the planned second burn was not performed. The command module landed within 50 miles of the onboard targeted landing point and was recovered in good condition by USS Okinawa.

APOLLO LTA-8
(Lunar Module Test Article #8) SUMMARY

GENERAL

Spacecraft: LTA-8
First Flight Crew:
Jim Irwin
Gerald Gibbons
Lockdown Time:
May 27th 1968
Mission Duration: 12 hrs
Lockdown Time:
May 29th 1968
Mission Duration: 12 hrs
Lockdown Time:
June 1st 1968
Mission Duration: 12 hours
Second Flight Crew:
Joseph Gagliano
Glennon Kingsley
Lockdown Time: May 31st 1968
Mission Duration: 10 hours
Third Flight Crew: Jim Irwin
Gerald Gibbons
Lockdown Time: Nov 14th 1968
Mission Duration: 13 hours

MISSION PRIMARY OBJECTIVES

1. To check out the spacecraft environmental control system
2. Proving the spacecraft structure and pressure vessel

DETAILED OBJECTIVES & EXPERIMENTS

1. Activating and checking out systems
2. Exercising computers
3. Computer Maneuvers
4. Simulated engine firings

UNUSUAL FEATURES OF THE MISSION

Mission LTA-8 took place inside the Manned Spacecraft Centre's Space Environment Simulation Laboratory, Chamber B. A stainless steel tank which can simulate the vacuum and temperatures of space.

REMARKS

Original crew member John Bull was taken off the crew roster almost immediately due to disorientation from an ear disorder. Most Mission objectives were successful. Mission LTA-8 was divided into two phases, each phase consisting of two mannings. The first test phase simulated temperatures expected on an earth orbital flight with the LM receiving minimum heating from the sun. The second test phase simulated maximum solar heating. During the tests the chamber maintained a vacuum simulating an altitude of up to 133 nautical miles and the walls of the chamber were cooled with liquid nitrogen to a temperature of about −300°F. Strip heaters attached to the skin of the LM supplied solar effects and soakback effects from engine firings. A scheduled test of the PLSS backpack was scrubbed due to an inadequate communications and data flow with the unit. The mission marked the first time a crew had egressed and ingressed a spacecraft in the facility at hard vacuum and included the first unscheduled repair of a spacecraft under vacuum conditions by crewmen in pressurized suits. The repair took place during the fourth manning when Irwin and Gibbons had to repair the hinge pins on the hatch of LTA-8. The pins were deliberately weak as a safety precaution for ground tests.

APOLLO 2TV-1
(Block 2 Thermal Vacuum test #1) SUMMARY

GENERAL
Spacecraft: CM-2TV-1
First Crew:
Joe Engle
Vance Brand
Joseph Kerwin
Lockdown Time: 2.23 pm June 16th 1968
Mission Duration: 7 days 16 hrs 7 mins
Mission end: 6.23 am June 24th 1968
Second Crew: Turnage Lindsey
Lloyd Reeder, Alfred Davidson
Lockdown Time: 1.00 pm August 9th 1968
Mission Duration: 16 hours
Mission end: 5.00 am August 10th 1968
Lockdown Time: Sept. 4th 1968
Mission Duration: 5 days
Mission end: 4.00 pm September 9th 1968

MISSION PRIMARY OBJECTIVES
1. To check out the spacecraft environmental control system
2. Proving the spacecraft structure and pressure vessel

DETAILED OBJECTIVES & EXPERIMENTS
1. Operating guidance and navigation equipment
2. Simulating engine firings
3. Activating and checking out spacecraft systems
4. Testing hatches
5. Testing new model Apollo pressure suits

UNUSUAL FEATURES OF THE MISSION

Mission 2TV-1 took place inside the Manned Spacecraft Centre's Space Environment Simulation Laboratory, Chamber A. A 120 foot high, 65 foot diameter stainless steel tank which can simulate the vacuum and temperatures of space. The first mission included six major phases. Crew ingress and a 19 hour chamber pumpdown phase, a 15 hour hot soak phase with the command module oriented toward the solar simulators followed by a 15 hour cold soak phase with solar simulators off. 45 hours with the chamber side solar simulators providing maximum heating to the service module, 71 hours of alternate and contingency operations and a 12 hour entry phase. For one period the solar simulators were inoperative so the timeline was revised to compensate. The second mission for 2TV-1 in August was primarily to verify test procedures and spacecraft systems. The crew wore the latest model of Apollo pressure suits. The 2TV-1 vehicle had undergone modifications to the forward and side hatches and the forward heat shield and had been equipped with a docking probe after completing the tests in June. The third mission for 2TV-1 included a cabin depressurization during the second day and the crew opened the hatch simulating an EVA. Some two and a half hours later they closed the hatch and repressurized the vehicle without difficulty. The crew remained out of their pressure suits and in a shirt sleeve environment during most of the five day test. Liquid nitrogen was used to cool the spacecraft walls to simulate the cold of space.

REMARKS

All Mission objectives were successful. Mission 2TV-1 subjected the spacecraft to temperatures ranging from −150°F to +150°F. The spacecraft was built to the same exacting specifications of the same materials and with nearly all of the same flight qualified equipment aboard as the Apolo spacecraft which will fly in space. It also includes a new quick release hatch and fire-proof cabin materials. The crew carried out most of the same functions as a crew would in space including eating and sleeping. The command module was filled with a mixture of 60% oxygen and 40% nitrogen atmosphere until it was gradually replaced with a pure 100% oxygen atmosphere at five pounds per square inch pressure. The spacecraft was mounted vertically on a rotating platform so it can be exposed to a wide range of simulated solar effects.

Astronaut Joe Engle climbs into the 2TV-1 command module. Note the non-standard cloth covered Apollo helmet.

APOLLO 7 (AS-205)
FLIGHT SUMMARY

GENERAL
Spacecraft: CM-101,
SM-101
Launch Vehicle: SA-205
Launch Complex: 34
Flight Crew:
Commander   (CDR)
Walter M. Schirra, Jr.
Command Module Pilot
(CMP) Donn F. Eisele
Lunar Module Pilot (LMP)
Walter Cunningham
Launch Time: 11:02:45
a.m. EDT, October 11,
1968
Launch Azimuth: 72°
Apogee : 245 NM
Perigee : 90 NM
Revolutions: 163
Mission Duration:
10 days 20 hours
Time of Landing: 7:11 a.m.
EDT, October 22, 1968

SPACE VEHICLE AND
PRE-LAUNCH DATA

APOLLO 7

Spacecraft delivered to
KSC: May 1968
Launch vehicle delivered to Cape
Kennedy:
First stage (S-IB): March 1968
Second stage (S-IVB): April 1968
Instrument Unit (IU): April 1968
Spacecraft weight at liftoff:
45,374 lb.
Space vehicle weight at liftoff:
1,277,742 lb.

MISSION PRIMARY
OBJECTIVES
(All Accomplished)

1 Demonstrate CSM/
crew performance.
2. Demonstrate
crew/space vehicle/
mission support facilities
performance during a
manned CSM mission.
3. Demonstrate CSM
rendezvous capability.

DETAILED TEST
OBJECTIVES
PRINCIPAL AND
MANDATORY
OBJECTIVES

Launch Vehicle:
(All Accomplished)
1. Demonstrate orbital
safing of the S-IVB.
2. Demonstrate launch
vehicle attitude control.
3. Qualify J-2 engine aug-
mented spark ignition
(ASI) line modification.

Spacecraft:
(All Accomplished)
1. Obtain data on the environmental
control system primary radiator
thermal coating degradation.
2. Obtain data on the Block II forward
heat shield thermal protection system.
3. Perform an inertial measurement
unit orientation determination and a
star pattern daylight visibility check.
4. Perform inertial measurement unit
alignments using the Sextant.
5. Perform guidance navigation control System controlled SPS and RCS
velocity maneuvers.
6. Demonstrate guidance navigation

control system automatic and manual attitude-controlled RCS maneuvers.

7. Evaluate the ability of the guidance navigation control system to guide the entry from earth orbit.

8. Demonstrate the stabilization control system automatic and manual attitude- controlled RCS maneuvers.

9. Demonstrate CSM stabilization control system velocity control capability.

10. Verify the life support functions of the environmental control system throughout the mission.

11. Demonstrate the water management subsystems operation in the flight environment.

12. Monitor the entry monitoring system during SPS velocity changes and entry.

13. Perform star and earth horizon sightings to establish an earth horizon model.

14. Obtain data on all command/service module consumables.

15. Demonstrate fuel cell water operations in a zero-g environment.

16. Perform a service propulsion system performance burn in the space environment.

17. Demonstrate the Performance of the command/service module - Manned Space Flight Network S-band communication system.

18. Verify the adequacy of the propellant feed line thermal control system.

19. Obtain inertial measurement unit Performance data in the flight environment.

20. Demonstrate the service propulsion system minimum impulse burns in a space environment.

21. Perform onboard navigation using the technique of the scanning telescope landmark tracking.

22. Obtain data on the stabilization control systems capability to provide a suitable inertial reference in a flight environment.

23. Verify automatic pressure control of the cryogenic tank systems in a zero-g environment.

24. Obtain data on thermal stratification with and without the cryogenic fans of the cryogenic gas storage system.

25. Demonstrate S-band updata link capability.

26. Obtain crew evaluation of intravehicular activity in general.

27. Obtain data on operation of the waste management system in the flight environment.

28. Operate the secondary coolant loop.

29. Perform a command/service module-active rendezvous with the S-IVB.

30. Accomplish the backup mode of the gyro display. coupler-flight director attitude indicator alignment using the Scanning telescope in preparation for an incremental velocity maneuver.

31. Demonstrate the postlanding ven-

Apollo 7 is prepared for launch, 1968.

tilation circuit operation.

32. Perform optical tracking of a target vehicle using the sextant.

33. Perform a command/service module - S-IVB separation, transposition and simulated docking.

34. Perform a manual thrust vector control takeover.

35. Monitor the primary and auxiliary gauging system.

36. Demonstrate a simulated command/service module overpass of the lunar module rendezvous radar during the lunar stay.

SECONDARY OBJECTIVES
Launch Vehicle:

1. Evaluate launch vehicle orbital lifetime. (Accomplished)

2. Demonstrate CSM manual launch vehicle orbital attitude control. (Accomplished)

Spacecraft: (All Accomplished)

1. Obtain data on initial coning angles when in the spin mode as used during transearth flight.

2. Demonstrate command/service module VHF voice communications with the Manned Space Flight Network.

3. Obtain data on the service module reaction control subsystem pulse and steady state performance.

4. Obtain data on propellant slosh damping following SPS cutoff and following reaction control subsystem burns.

5. Verify that the launch vehicle propellant pressure displays are adequate to warn of a common bulkhead reversal.

6. Obtain photographs of the command module rendezvous windows during discrete phases of the flight.

7. Evaluate the crew optical alignment

sight for docking, rendezvous and proper attitude verification

8. Perform manual out-of-window command/service module attitude orientation for retro fire.

9. Monitor the guidance navigation control systems and displays during launch.

10. Obtain data via the command/service module - Apollo Range Instrumentation Aircraft communications systems.

11. Perform Crew-controlled manual S-IVB attitude maneuvers in three axes.

12. Obtain data on the spacecraft-LM adapter deployment system operation.

13. Obtain command/service module vibration data.

14. Obtain selective, high quality photographs with color and panchromatic film of selected land and ocean areas.

15. Obtain selective, high quality, color cloud photographs to study the fine structure of the earth's weather system.

UNUSUAL FEATURES OF THE MISSION

1. First manned Apollo flight.

2. First flight of Block II Apollo Spacecraft.

3. First flight of the Apollo space suits.

4. First flight with full crew support equipment.

5. First live national TV from space during a manned space flight.

Significant Spacecraft changes from Block I:

* A unified hatch assembly was incorporated.

* S-band equipment was added.

* Unitized crew couches were incor-

porated.

* Flight qualification and operational instrumentation were increased.
* Full crew support systems were incorporated.
* Usage of non-metallic materials was modified and decreased.
* A 60% oxygen/40% nitrogen cabin environment was used during pre-launch and early boost phases of the mission.
* There was an increased use of stainless steel tubing in place of aluminum.
* Armoring of solder tubing joints was increased.
* Fire extinguisher and emergency oxygen masks were incorporated in the CM.
* An onboard TV camera was added.
* The capabilities of components of the earth landing system were improved.
* Communication system modifications were incorporated.
* A redesigned cobra cable was incorporated.

RECOVERY DATA

Recovery Area:
West Atlantic Ocean
Landing Coordinates: 27°33'N., 64°04'W. (Stable II)
Recovery Ship: USS Essex
Crew Recovery Time: 8:08 a.m. EDT, October 22, 1968
Spacecraft Recovery Time: 9:03 a.m. EDT, October 22, 1968

REMARKS

All primary Apollo 7 Mission objectives were successfully accomplished. In addition, all planned detailed test objectives plus three that were not originally scheduled were satisfactorily accomplished. As part of the effort to alleviate fire hazard prior to liftoff and during initial flight, the command module cabin atmosphere was composed of 60% oxygen and 40% nitrogen. During this period the crew was isolated from the cabin by the suit circuit, which contained 100% oxygen. Shortly after liftoff, the cabin atmosphere was gradually enriched to pure oxygen at a pressure of 5 psi. Hot meals and relatively complete freedom of motion in the spacecraft enhanced crew comfort over previous Mercury and Gemini flights. The service module SPS main engine proved itself by accomplishing the longest and shortest manned SPS burns and the largest number of inflight restarts. The SPS engine is the largest thrust engine to be manually thrust vector-controlled. Manual tracking, navigation, and control achievements included full optical rendezvous, daylight platform realignment, optical platform alignments, pilot attitude control of launch vehicle, and orbital determination by sextant tracking of another vehicle by the spacecraft. The Apollo 7 Mission also accomplished the first digital auto pilot-controlled engine burn and the first manned S-band communications. All launch vehicle systems performed satisfactorily throughout their expected lifetime. All spacecraft systems continued to function throughout the mission with some minor anomalies. Each anomaly was countered by a backup subsystem, a change in procedures, isolation, or careful monitoring such that no loss of system support resulted. Temperatures and consumables usages remained within specified limits throughout the mission.

APOLLO 8 (AS-503)
FLIGHT SUMMARY

GENERAL

Spacecraft: CM-103, SM-103, LTA-B
Launch Vehicle: SA-503
Launch Complex: 39A
Flight Crew:
   Commander (CDR) Frank Borman
   Command Module Pilot (CMP) James A. Lovell, Jr.
   Lunar Module Pilot (LMP) William A. Anders
Launch Time: 7:51:00 a.m. EST, December 21, 1968
Launch Azimuth: 72°
Earth Orbit: Apogee 103.3 NM, Perigee 98.0 NM 54
Lunar Orbit:
   Initial Apocynthion 168.5 NM, Pericynthion 59.7 NM
   Circularized Apocynthion 60.7NM, Pericynthion 59.7 NM
Mission Duration: 146 hours 59 minutes 49 seconds
Time of Landing: 10:50:49 a.m. EST, December 27, 1968

SPACE VEHICLE AND PRE-LAUNCH DATA

Spacecraft delivered to KSC:
   Command/service module (CSM): August 1968

APOLLO 8

Lunar module test article (LTA) : January 1968
Launch vehicle delivered to KSC:
First stage (S-IC): December 1967
Second stage (S-II): June 1968
Third stage (S-IVB): December 1967
Instrument unit (IU): January 1968
Space vehicle weight at liftoff: 6,133,880 lb.
Weight placed in earth orbit: 282,237 lb.
Weight placed in lunar orbit: 46,743 lb.

MISSION PRIMARY OBJECTIVES
(All Accomplished)

1. Demonstrate Crew/space vehicle/mission support facilities performance during a manned Saturn V mission with CSM.
2. Demonstrate performance of nominal and selected backup lunar orbit rendezvous (LOR) mission activities, including:
a. Translunar injection;
b. CSM navigation, communications, and midcourse corrections;
c. CSM consumables assessment and passive thermal control.

DETAILED TEST OBJECTIVES PRINCIPAL AND MANDATORY OBJECTIVES

Launch Vehicle:
(All Accomplished)

1. Verify the capability of the launch vehicle to perform a free-return translunar injection (TLI).

2. Demonstrate the capability of the S-IVB to restart in earth orbit.

3. Verify the modifications made to the J-2 engine since the Apollo 6 Flight.

4. Confirm the J-2 engine environment in the S-II and S-IVB stages.

5. Confirm the launch vehicle longitudinal oscillation environment during the S-IC stage burn.

6. Verify that the modifications incorporated in the S-IC stage since the Apollo 6 flight suppress low frequency longitudinal oscillations (POGO).

7. Demonstrate the operation of the S-IVB helium heater repressurization system.

8. Verify the capability to inject the S-IVB/IU/LTA-B into a lunar "slingshot" trajectory.

9. Demonstrate the capability to safe the S-IVB stage in orbit.

10. Verify the onboard command and communication system (CCS) and ground system interface and the operation of the CCS in a deep space environment.

Spacecraft:

1. Perform a guidance, navigation, and control system (GNCS)-controlled entry from a lunar return. (Accomplished)

2. Perform star-lunar horizon sightings during the translunar and transearth phases. (Accomplished)

3. Perform star-earth horizon sightings during translunar and transearth phases. (Accomplished)

4. Perform manual and automatic acquisition, tracking, and communication with MSFN using the high-gain CSM S-band antenna during a lunar mission. (Accomplished)

5. Obtain data on the passive thermal control system during a lunar orbit mission. (Accomplished.)

6. Obtain data on the spacecraft dynamic response. (Accomplished)

7. Demonstrate SLA panel jettison in a zero-g environment. (Accomplished)

8. Perform lunar orbit insertion SPS GNCS- controlled burns with a fully loaded CSM. (Accomplished)

9. Perform a transearth insertion GNCS- controlled SPS burn. (Accomplished)

10. Obtain data on the CM Crew procedures and timeline for lunar orbit mission activities. (Accomplished)

11. Demonstrate CSM passive thermal control (PTC) modes and related communication procedures during a lunar orbit mission. (Accomplished)

12. Demonstrate ground operational support for a CSM lunar orbit mission. (Accomplished)

13. Perform lunar landmark tracking from the CSM in lunar orbit. (The intent of this objective was to establish that an onboard capability existed to compute relative position data for the lunar landing mission. This mode will be used in conjunction with the MSFC state-vector update.) (Partially Accomplished)

14. Prepare for translunar injection (TLI), and monitor the GNCS and LV tank pressure displays during the TLI burn. (Accomplished)

15. Perform translunar and transearth midcourse corrections. (Accomplished)

SECONDARY OBJECTIVES

Spacecraft:

1. Monitor the GNCS and displays during launch. (Accomplished)

2. Obtain IMU performance data in the flight environment. (Accomplished)

3. Perform star-earth landmark sighting navigation during translunar and transearth phases. (The intent of this objective was to demonstrate onboard star-earth landmark optical navigation.) (Partially Accomplished)

4. Perform an IMU alignment and a star pattern visibility check in daylight. (Accomplished)

5. Perform SPS lunar orbit insertion and transearth injection burns and monitor the primary and auxiliary gauging systems. (Accomplished)

6. Obtain data on the Block II ECS performance during manned lunar return entry conditions. (Accomplished)

7. Communicate with MSFC using the CSM S-band omni-antennas at lunar distance. (Accomplished)

8. Demonstrate the performance of the Block II thermal protection system during a manned lunar return entry. (Accomplished)

9. Perform a CSM/S-IVB separation and a CSM transposition on a lunar mission timeline. (Accomplished)

10. Obtain data on CSM consumables for a CSM lunar orbit mission. (Accomplished)

11. Obtain photographs during the transearth, translunar and lunar orbit phases for operational and scientific purposes. (Accomplished)

12. Obtain data to determine the effect of the tower jettison motor, S-II retro and SM RCS exhausts and other sources of contamination on the CM windows. (Accomplished)

UNUSUAL FEATURES OF THE MISSION

1. First manned Saturn V flight.

2. First manned flight to the lunar vicinity.

3. Highest velocity yet attained by man - 36,228 fps.

4. First live TV coverage of the lunar surface.

5. Deepest penetration of space by a manned spacecraft.

6. First space flight on which man escaped earth's gravity.

Significant spacecraft differences from Apollo 7:

* Forward hatch was modified to a combined forward Crew hatch.

* The SM aft bulkhead structure was modified to assure a 1.4 factor of safety.

* The CM-SM tension tie thickness was increased.

* The SM/SLA interface was redesigned to install bolts from outside.

* Couch strut load/stroke criteria were reduced and lockouts added.

* A change to foldable crew couches was incorporated.

* The spacecraft ground intercom was converted from a two-wire to a four- wire system.

* An S-band high-gain antenna was included.

* A high-gain antenna automatic reacquisition system was added.

* The ECS radiator flow proportioning valve was redesigned.

* Aluminum C02 absorber elements were added.

* The Colossus onboard software was installed.

* A change to jettisonable SLA panels

was incorporated.
* The Van Allen Belt dosimeter was added.
* POGO instrumentation was added.
* A nuclear particle detection system was added.
* The right-hand crewman's right-hand arm rest was deleted.
* A redundant launch vehicle attitude error display was added. Significant launch vehicle changes from Apollo 6:
* The ASI's in the J-2 engine were modified.
* The S-IC stage was modified to suppress low frequency longitudinal oscillations.

RECOVERY DATA
Recovery Area: Pacific Ocean
Landing Coordinates: 165°1'W. 8°8'N. (Stable II)
Recovery Ship: USS Yorktown
Crew
Recovery Time: 12:20 p.m. EST, December 27, 1968
Spacecraft Recovery Time: 13:20 p.m. EST, December 27, 1968

REMARKS
All primary Apollo 8 mission objectives were completely accomplished. Every detailed test objective was accomplished as well as four which were not originally planned. The AS-503 Space Vehicle featured several configuration details for the first time, including: a Block II Apollo Spacecraft on a Saturn V Launch Vehicle, a manned spacecraft on a Saturn V Launch Vehicle, an O2H2 gas burner on the S-IVB for propellant tank repressurization Prior to engine restart, open-loop propellant utilization

systems on the S-II and S-IVB stages, and jettisonable SLA Panels. For this first Apollo flight to the lunar vicinity, Mission Operations successfully coped with lunar launch opportunity and launch window constraints and injected the S-IVB into a lunar "slingshot" trajectory to prevent recontact with the spacecraft or impact on the moon or earth. Apollo 8 provided man his first opportunity to personally view the backside of the moon, view the moon from as little as 60 NM away, view the earth from over 200,000 NM away, and reenter the earth's atmosphere through a lunar return corridor at lunar return velocity. All launch vehicle systems performed satisfactorily throughout their expected lifetimes. All spacecraft systems continued to function satisfactorily throughout the mission. No major anomalies occurred. Those minor discrepancies which did occur were primarily procedural and were corrected in flight with no mission impact. All temperatures and consumables usage rates remained within normal limits throughout the mission.

Nestled inside the S-IVB stage is the Apollo 8 LTA-B

APOLLO 9 (AS-504)
FLIGHT SUMMARY

GENERAL

Spacecraft: CM-104, SM-104, LM- 3
Launch Vehicle: SA-504
Launch Complex: 39A
Flight Crew:
Commander (CDR) James A. McDivitt
Command Module Pilot (CMP) David R. Scott
Lunar Module Pilot (LMP) Russell L. Schweickart
Launch Time: 11:00:00 a.m. EST, March 3, 1969
Launch Azimuth: 72°
Apogee: 271.8 NM (Highest)
Perigee: 97.8 NM (Lowest)
Mission Duration: 10 days 01 hour 53 seconds
Time of Landing: 12:00:53 p.m. EST, March 13, 1969

SPACE VEHICLE AND PRE-LAUNCH DATA

Spacecraft delivered to KSC:
Command/service module (CSM) : October 1968
Lunar module (LM): June 1968
Launch vehicle delivered to KSC:
First stage (S-IC) :

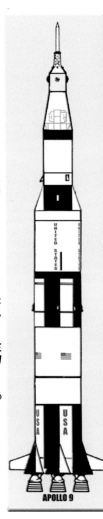

APOLLO 9

September 1968
Second stage (S-II): May 1968
Third stage (S-IVB): September 1968
Instrument unit (IU): September 1968
Space vehicle weight at liftoff: 6,397,055 lb.
Weight placed in earth orbit: 292,091 lb.

MISSION PRIMARY OBJECTIVES
(All Accomplished)

1. Demonstrate crew/space vehicle/mission support facilities performance during a manned Saturn V mission with CSM and LM.
2. Demonstrate LM/crew performance.
3. Demonstrate performance of nominal and selected backup LOR mission activities, including:
a. Transposition, docking, LM withdrawal;
b. Intervehicular crew transfer;
c. EVA capability;
d. SPS and DPS burns;
e. LM-active rendezvous and docking.
4. CSM/LM consumables assessment.

DETAILED TEST OBJECTIVES PRINCIPAL AND MANDATORY OBJECTIVES

Launch Vehicle:

1. Demonstrate S-IVB/IU attitude control capability during transposition, docking, and LM ejection (TD&E) maneuver. (Accomplished.)

Spacecraft:

(All Accomplished except where noted)

1. Perform LM-active rendezvous.
2. Determine DPS duration effects and primary propulsion/ vehicle interactions.
3. Verify satisfactory performance of passive thermal sub-system.
4. Demonstrate LM structural integrity.
5. Perform DPS burn including throttling, docked; and a short duration DPS burn, undocked.
6. Perform long duration APS burns.
7. Demonstrate environmental control system (ECS) performance during all LM activities.
8. Obtain temperature data on deployed landing gear resulting from DPS operation.
9. Determine electrical power system (EPS) performance, primary and backup.
10. Operate landing radar during DPS burns.
11. Perform abort guidance system (AGS)/control electronics system (CES)-controlled DPS burn.
12. Perform primary guidance, navigation, and control system (PGNCS)/digital auto pilot (DAP)-controlled long duration APS burn.
13. Demonstrate RCS control of LM using manual and automatic PGNCS.
14. Demonstrate S-band and VHF communication compatibility. (Partially Accomplished)
15. Demonstrate RCS control of LM using manual and automatic AGS/CES.
16. Demonstrate CSM attitude control, docked, during SPS burn.
17. Demonstrate LM-active docking.
18. Demonstrate LM ejection from SLA with CSM.
19. Demonstrate CSM-active docking.
20. Demonstrate CSM-active undocking.
21. Verify inertial measurement unit (IMU) performance.
22. Demonstrate guidance, navigation, and control system (GNCS)/manual thrust vector control (MTVC) takeover.
23. Demonstrate LM rendezvous radar performance.
24. Demonstrate LM/Manned Space Flight Network (MSFN) S-band communications capability. (Partially Accomplished)
25. Demonstrate intervehicular transfer (IVT).
26. Demonstrate AGS calibration and obtain performance data in flight.
27. Perform LM IMU alignment.
28. Perform LM jettison.
29. Obtain data on reaction control system (RCS) plume impingement and corona effect on rendezvous radar performance.
30. Demonstrate support facilities performance during earth orbital missions.
31. Perform IMU alignment and daylight star visibility check, docked.
32. Prepare for CSM-active rendezvous with LM.
33. Perform IMU alignment with sextant (SXT), docked.
34. Perform landing radar self-test.
35. Perform extravehicular activity.

SECONDARY OBJECTIVES

Launch Vehicle:

(All Accomplished)

1. Verify S-IVB restart capability.

2. Verify J-2 engine modification.

3. Confirm J-2 engine environment in S-II stage.

4. Confirm launch vehicle longitudinal oscillation environment during S-IC stage burn period.

5. Demonstrate O2H2 burner repressurization system operation.

6. Demonstrate S-IVB propellant dump and safing. (Not Accomplished)

7. Verify that modifications incorporated in the S-IC stage suppress low-frequency longitudinal oscillations.

8. Demonstrate 80-minute restart capability.

9. Demonstrate dual repressurization capability.

10. Demonstrate O2H2 burner restart capability.

11. Verify the onboard command and communications system (CCS)/ground system interface and operation in the space environment.

Spacecraft:

(All Accomplished)

1. Obtain exhaust effects data from launch escape tower (LET), S-II retro, and SM RCS on CSM.

2. Evaluate Crew performance of all tasks.

3. Perform navigation by landmark tracking.

4. Perform unmanned APS burn-to-depletion.

5. Obtain data on DPS plume effects on visibility.

6. Perform CSM/LM electromagnetic compatibility test.

UNUSUAL FEATURES OF THE MISSION

1. Largest payload yet placed in orbit.

2. First launch of Saturn V/Apollo Spacecraft in lunar mission configuration.

3. First demonstration of S-IVB second orbital restart capability.

4. First CSM-active docking.

5. First manned LM systems performance demonstration.

6. First inflight depressurization and hatch opening of LM and CM.

7. First Apollo extravehicular activity.

8. First intervehicular transfer between docked interface of two vehicles in shirt sleeve environment.

9. First docked SPS burns with CSM guidance and docked DPS burns with LM guidance.

10. First demonstration of lunar module TV camera (black and white).

11. First LM TV.

12. First LM-active rendezvous and docking.

13. First time one spacecraft was configured from another spacecraft for an unmanned burn.

Significant spacecraft differences from Apollo 8 (LM-3 as compared with LM-1 which was flown on Apollo 5):

Command Module

* Forward hatch emergency closing link was added.

* A general purpose timer was added.

* A precured RTV was added to side and hatch windows.

* The S-065 camera experiment equipment was added.

* Docking probe, ring, and latches were added.

* An RCS propulsion burst disc was added.

* A solenoid valve was added to the RCS propellant system.

* The S-band power amplifier configuration was changed to 0006 con-

figuration.

* The flight qualification recorder was deleted.

Lunar Module

* First operational flight of oxygen supply module.
* First operational flight of water control module.
* First flight of VHF transceiver and diplexer.
* First flight to use exterior tracking light.
* First flight to use ascent engine arming assembly.
* First operational flight of the abort guidance section.
* First operational flight of the rendezvous radar.
* First flight of the landing radar electronic and antenna assembly.
* First flight using thrust translation controller assembly.
* First flight to use orbital rate drive.
* The CO_2 partial pressure sensor was modified to correct EMI, vibration, and outgassing problems.
* A high-reliability transformer was added for use with the S-band steerable antenna.
* A pressure switch was added to the RCS.
* Thermal insulation was modified in the rendezvous radar antenna assembly.
* Landing gear was installed.
* High-efficiency reflective coated cabin and docking windows were added.
* A split AC bus was added.
* A more reliable signal processor assembly was added.
* Manual trim shutdown was added to descent engine control assembly.
* Stabilization and control assembly No. 1 was modified to eliminate single failure point.

* Fire preventive and resistive materials were added.
* A TV camera was added. Spacecraft-LM Adapter
* The SLA panel charges were redesigned.
* A spring ejector for LM separation was added.
* The LM separation sequence controllers were added.
* The POGO instrumentation was deleted.

Significant launch vehicle changes from Apollo 8:

S-IC Stage

* The film camera system was deleted.
* The R&D instrumentation was reduced.
* A redesigned F-1 engine injector was installed.
* Television cameras were removed.
* Propulsion performance was increased.
* Weight was reduced by removal of forward skirt insulation and revising "Y" rings and skin taper in propellant tanks.

S-II Stage

* First flight of lightweight structure.
* Separation planes tension plates were redesigned.
* The J-2 engines were uprated.
* The thrust structure was reinforced.
* The propellant utilization (PU) system was changed to closed loop.

S-IVB Stage

* Instrumentation battery capacity was reduced.
* The anti-flutter kit was deleted.
* The J-2 engine was uprated.

Instrument Unit
* The methanol accumulator was enlarged.
* Networks to disable spacecraft control of launch vehicle were changed.
* One instrument battery was removed.
* The S-band telemetry was deleted.

RECOVERY DATA
Recovery Area:
Atlantic Ocean Landing
Coordinates:
67°56'W., 23°13'N. (Stable I)
Recovery Ship:
USS Guadalcanal Crew
Recovery Time:
12:50 p.m. EST, March 13, 1969
Spacecraft Recovery Time:
2:13 p.m. EST, March 13, 1969

REMARKS
A mild virus respiratory illness which infected all of the Apollo 9 Crew members was the primary factor in the decision to reschedule the launch from February 28 to 11:00 EST, March 3, 1969. This decision to reschedule was made February 27, 1969 in order to assure the full recovery and good health of the astronauts. The Countdown was accomplished without any unscheduled holds and was the fourth Saturn V on-time launch. The Apollo 9 launch was the First Saturn V/Apollo Spacecraft in full lunar mission configuration and carried the largest payload ever placed in orbit. Since Apollo 9 was the first manned demonstration of lunar module systems performance, many firsts were achieved. These were highlighted by CSM and LM-active rendezvous and docking, the First Apollo EVA, and intervehicular transfer in shirt sleeve environment. This flight also contained the first demonstration of S-IVB second orbital restart capability. In the third day of the mission, LMP Schweickart was struck by nausea and this illness caused a Small delay from the normal timeline in the donning of pressure suits and in the transfer to the LM. It also caused shortening of the proposed EVA plan. Later the next morning, CDR McDivitt assessed LMP Schweickart's condition as excellent and with ground control concurrence decided to extend his EVA activities. The Apollo 9 crew had remarkable success in sighting objects using the crewman optical alignment sight (COAS). Their success seems to confirm the thesis that the visual acuity of the human eye is increased in space. One example is their sighting of the Pegasus II Satellite at a range of approximately 1,000 miles. All primary objectives were successfully accomplished on the Apollo 9 flight. All mandatory and principal detailed test objectives were accomplished, except two, and these two were partially accomplished. One secondary detailed test objective, the S-IVB propellant dump and safing, was not accomplished. All launch vehicle systems performed satisfactorily throughout their expected lifetimes with the exception of inability to dump propellants following the third S-IVB burn. All spacecraft systems continued to function satisfactorily throughout the mission. No major anomalies occurred. Those minor discrepancies which did occur were primarily procedural and were. corrected in flight with no mission impact, or involved instrumentation errors in quantities which could be checked by other means. Temperatures and consumables usage rates remained generally within normal limits throughout the mission.

APOLLO 10 (AS-505)
FLIGHT SUMMARY

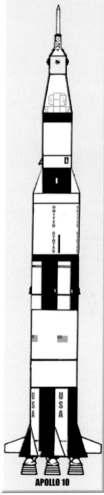

APOLLO 10

GENERAL
Spacecraft: CM-106, SM-106, LM-4
Launch Vehicle: SA-505
Launch Complex: 39B
Flight Crew:
  Commander (CDR)
  Thomas P. Stafford
  Command Module Pilot (CMP) John W. Young
  Lunar Module Pilot (LMP) Eugene A. Cernan
Launch Time: 12:49 p.m. EDT, May 18, 1969
Launch Azimuth: 72 °
Earth Orbit:
Apogee: 102.6 NM
Perigee: 99.6 NM
Lunar Orbits:
Initial Apocynthion/Pericynthion (LOI-1): 170.4 NM x 59.6 NM
Circularized Apocynthion/Pericynthion (LOI- 2): 61.5 NM x 58.9 NM
LM Descent Orbit Insertion: 61.2 x 8.4 NM
LM Phasing Maneuver: 190 NM x 11.2 NM
LM Insertion Maneuver: 45.3 NM x 11.2 NM
Final LM/CSM Separation: 63.2 NM x 55 NM
Mission Duration: 192 hours 3 minutes 23 seconds
Time of Landing: 12:52:23 EDT, May 26, 1969

SPACE VEHICLE AND PRE-LAUNCH DATA
Spacecraft delivered to KSC:
Command/service module (CSM): November 1968
Lunar module (LM): October 1968
Launch vehicle delivered to KSC:
First stage (S-IC): November 1968
Second stage (S-II): December 1968
Third stage (S-IVB): December 1968
Instrument unit (IU): December 1968
Space vehicle weight at liftoff: 6,412,250 lb.
Weight placed in earth orbit: 294,947 lb.
Weight placed in lunar orbit: 69,429 lb.

MISSION PRIMARY OBJECTIVES
(All Accomplished)

1. Demonstrate crew/space vehicle/mission support facilities performance during a manned lunar mission with CSM and LM.
2. Evaluate LM performance in the cislunar and lunar environment.

DETAILED TEST OBJECTIVES PRINCIPAL AND MANDATORY OBJECTIVES

Spacecraft
(All Accomplished)
1. Demonstrate CSM/LM rendezvous capability for a lunar landing mission.
2. Perform manual and automatic acquisition, tracking, and communications with MSFN using the steerable S-band antenna at lunar distance.
3. Perform lunar landmark tracking from the CSM while in lunar orbit.
4. Perform lunar landmark tracking in lunar orbit from the CSM with the LM attached.
5. Operate the landing radar at the closest approach to the moon and during DPS burns.
6. Obtain data on the CM and LM crew procedures and timeline for the lunar orbit phase of a lunar landing mission.
7. Perform PGNCS/DPS undocked descent orbit insertion (DOI) and a .high thrust maneuver.

SECONDARY OBJECTIVES

Launch Vehicle:
(All Accomplished)
1. Verify J-2 engine modifications.
2. Confirm J-2 engine environment in S-II and S-IVB stages.
3. Confirm launch vehicle longitudinal oscillation environment during S-IC stage burn period.
4. Verify that modifications incorporated in the S-IC stage suppress low frequency longitudinal oscillations.
5. Confirm launch vehicle longitudinal oscillation environment during S-II stage burn period.
6. Demonstrate that early center engine cutoff for S-II stage suppresses low frequency longitudinal oscillations.

Spacecraft
(All Accomplished except where noted)
1. Demonstrate LM/CSM/MSFN communications at lunar distance. (Partially Accomplished)
2. Communicate with MSFN using the LM S-band omni antennas at lunar distance.
3. Obtain data on the rendezvous radar performance and capability near maximum range.
4. Obtain supercritical helium system pressure data while in standby conditions and during all DPS engine firings.
5. Perform an unmanned AGS-controlled APS burn.
6. Obtain data on the operational capability of VHF ranging during an LM-active rendezvous.
7. Obtain data on the effects of lunar illumination and contrast conditions on crew visual perception while in lunar orbit.
8. Obtain data on the passive thermal control mode during a lunar orbit mission. (Partially Accomplished)
9. Demonstrate CSM/LM passive thermal control modes during a lunar orbit mission.
10. Demonstrate RCS translation and attitude control of the staged LM using automatic and manual AGS/CES control.
11. Evaluate the ability of the AGS to perform an LM active rendezvous.
12. Monitor PGNCS/AGS performance during lunar orbit operations.
13. Demonstrate operational support for a CSM/LM lunar orbit mission.

14. Perform a long duration unmanned APS burn.

15. Perform lunar orbit insertion using SPS GNCS-controlled burns with a docked CSM/LM.

16. Obtain data to verify IMU performance in the flight environment.

17. Perform a reflectivity test using the CSM S-band high-gain antenna while docked.

18. Perform CSM transposition, docking, and CSM/LM ejection after S-IVB TLI burn.

19. Perform translunar midcourse corrections.

20. Obtain AGS performance data in the flight environment.

21. Perform star-lunar landmark sightings during the transearth phase.

22. Obtain data on LM consumables for a simulated lunar landing mission, in lunar orbit, to determine lunar landing mission consumables.

UNUSUAL FEATURES OF THE MISSION
Provided these first-time inflight opportunities:

1. Lunar orbit rendezvous.

2. Docked lunar landmark tracking.

3. Lunar module steerable antenna operation at distances greater than those of low earth orbit enabling its evaluation under conditions for which it was designed.

4. Descent propulsion system (DPS) engine burn in the lunar landing mission configuration and environment.

5. Lunar landing mission profile simulation (except for powered descent, lunar surface activity, and ascent).

6. LOW level (47,000 feet) evaluation of lunar visibility.

7. Docked CSM/LM thermal control

in the absence of earth albedo and during long periods of sunlight.

8. Lunar module omni-directional antenna operation at lunar distance.

9. Abort guidance system (AGS) operation during an APS burn over the range of inertias for a lunar mission.

10. VHF ranging during a rendezvous.

11. Landing radar operation near lunar environment where the reflected energy from the lunar surface is detected.

12. Transposition, docking, and LM ejection in daylight after the S-IVB burn where the S-IVB is in inertial hold attitude and while the spacecraft is moving away from the earth.

13. Translunar midcourse correction with a docked CSM/LM.

14. Lunar module digital uplink assembly first flight (replaces digital command assembly used on LM-3).

15. First launch from Pad B of launch complex 39.

16. Largest payload yet placed in earth orbit.

17. Largest payload yet placed in lunar orbit.

18. Demonstration of color TV camera.

19. Manned navigational, visual, and photographic evaluation of lunar landing sites 2 and 3.

20. Manned visual and photographic evaluation of range of possible landing sites in Apollo belt highlands areas.

21. Acquisition of major quantities of photographic training materials for Apollo 11 and subsequent lunar landing missions.

22. Acquisition of numerous visual observations and photographs of scientific significance.

Significant spacecraft differences from Apollo 9:

Command Module
* The VHF ranging capability was added as a Backup to CSM/LM rendezvous radar (RR).

Lunar Module
* The VHF ranging capability was added as an RR Backup.
*The CM to LM power transfer capability after LM stage separation was incorporated to extend hold capability between docking and final LM/CSM separation.
* The CM/LM power transfer redundancy was provided as a power transfer Backup.
* The EVA antenna was deleted because there was no EVA planned for Apollo 10.
* Digital uplink voice output (up to 20 db) was increased because it was required for lunar distance communication.
* Landing gear deployment mechanism protective shield was added to prevent possible malfunction due to DPS plume impingement.
*Ascent stage plume heat blanket and venting was added to improve thermal control.
* A separate power source for utility/floodlight was added to prevent simultaneous loss of both lights.
* An APS muffler was added to prevent APS regulator loss.
* RR and VHF bus isolation was provided to prevent simultaneous RR and VHF loss.
*The TV camera was deleted.
* Luminary 1 (LM onboard program) was used for the first time (Sundance for LM-3). Significant launch vehicle

changes from Apollo 9:

S-II Stage
* Center engine early cutoff was planned as a means of eliminating longitudinal oscillations.

S-IVB Stage
* A redesigned helium regulator valve was substituted to correct an SA-504 malfunction.

Instrument Unit
* Instrument unit network change (software) was incorporated to enable SC control of LV during the launch phase and translunar injection.
* Insulation and damping compound were added to improve vibration damping and IU load-carrying capability.

RECOVERY DATA
Recovery Area:
Southwest Pacific Ocean
Landing Coordinates:
15°S., 165°W. (Stable I)
Recovery Ship: USS Princeton Crew
Recovery Time: 1:31 p.m. EDT, May 26, 1969
Spacecraft Recovery Time: 2:22 p.m. EDT, May 26, 1969

REMARKS
The most complex mission yet flown in the Apollo Program was performed in the full lunar landing configuration, paralleling as closely as possible the lunar landing mission profile and timeline. Extensive photographic coverage of candidate lunar landing sites provided excellent data and crew training material for subsequent missions. This was the fifth on-time Saturn V launch. Nineteen color television transmis-

sions (totaling 5 hours 52 minutes) of remarkable quality provided a world audience the best exposure yet to spacecraft activities and spectacular views of the earth and the moon. The LM pericynthion of 47,000 feet was the closest man had come to the moon, and the crew reported excellent visual perception of the proposed landing areas. The mission was nominal in all major respects. Translunar and transearth navigational accuracy was so precise that only two of seven allocated midcourse corrections were required, one each during translunar and transearth coast periods. Significant perturbations in lunar orbit, resulting from differences in gravitational potential, were noted. Subsequent mission LOI burns can be biased to compensate for these effects. All launch vehicle systems performed satisfactorily during their expected lifetimes. Spacecraft systems generally performed satisfactorily throughout the mission. One exception was the No. I fuel cell which had to be isolated from the main bus, but work-around procedures made it available for load sharing, if required. Another problem was the occasional difficulty with direct LM-earth communications. Two incidents of unexpected motion occurred prior to and during LM staging. Data indicates unscheduled transfer of the abort guidance system mode from "Attitude Hold" to "Automatic." A number of minor discrepancies occurred which were either primarily procedural and were corrected in flight with no mission impact, or which involved instrumentation errors on quantities that could be checked by other means. Two cameras that malfunctioned were returned to earth for

failure analysis. All detailed test objectives were met, except for two secondary spacecraft objectives that were partially accomplished. Five other major activities not defined as detailed test objectives were fully accomplished. Flight crew performance was outstanding. Their health and spirits remained excellent throughout the mission. Unexpected bonuses from the mission were several sightings of individual SLA Panels long after TD&E, three sightings of the jettisoned descent stage as it orbited the moon at low altitude, and a few sightings of the receding S-IVB stage with the naked eye, once from nearly 4000 miles as it tumbled and flashed in the sunlight.

The first "Apollo" launch,
Little Joe II QTV
August 28th 1963

LITTLE JOE II & PAD ABORT TEST FLIGHTS

Item	QTV	A-001	A-002	A-003	A-004	LESPA-1	LESPA-2
Total	25,984 kg	26,339 kg	42,788 kg	80,372 kg	63,381 kg	9,800kg	9,800kg
Payload	11,011 kg	11,525 kg	12,561 kg	12,626 kg	14,717 kg	5,800kg	5,800kg
Airframe	14,973 kg	14,814 kg	29,320 kg	67,745 kg	48,623 kg		
Apogee	5.6 km	5.9 km	5 km	6 km	23.8 km	2.8 km	2km
Date	8/28/63	5/13/64	12/8/64	5/19/65	1/20/66	11/7/63	6/29/65

LJ-II QTV

LITTLE JOE QTV
TEST OBJECTIVES
The purpose of this test was to demonstrate the capability of the launch vehicle to perform adequately the launch phase of Apollo Mission A-001 with its particular rocket combination.
Demonstrate capability of the launch vehicle to clear the launcher at lift-off successfully.
Demonstrate capability of the launch vehicle to perform adequately the launch trajectory of Mission A-001.
Demonstrate the Algol motor thrust termination system.
Demonstrate the adequacy of the operational procedure of launcher elevation and azimuth setting to compensate for winds existing at time of launch.
Determine the vehicle base pressure.
Determine base heating.
Demonstrate that the launch vehicle fixed fins are flutter-free in the transonic region.
Demonstrate structural integrity of the launch vehicle to perform the Mission A-001.
Evaluate the techniques and procedures which contribute to efficient operations involving the launch of Apollo payloads on the Little Joe II launch vehicle.
Demonstrate the functional performance and structural adequacy of the ground support equipment.
Evaluate procedures for ground command abort to be used for Mission A-001.
Determine the overall vehicle flexible body response.

LJ-11 A001

LITTLE JOE A-001
TEST OBJECTIVES
Demonstrate the structural integrity of the escape tower.
Demonstrate the capability of the escape subsystem to propel the command module to a predetermined distance from the launch vehicle. Determine aerodynamic stability characteristics of the escape configuration for this abort condition.
Demonstrate proper operation of the command module to service module separation subsystem.
Demonstrate satisfactory recovery timing sequence in the earth-landing subsystem.
Demonstrate Little Joe II-spacecraft compatibility.
Determine aerodynamic loads due to fluctuating pressures on the command module and service module during a Little Joe II launch.

Demonstrate proper operation of the applicable components of the earth-landing subsystem.

LITTLE JOE A-002
TEST OBJECTIVES
Demonstrate satisfactory launch-escape vehicle power-on stability for abort in the maximum dynamic-pressure region which conditions approximating the attitude limit at which the emergency-detection system would trigger an abort.

LJ-II A002

Determine the command module (CM) pressure loads, including possible plume impingement, in the maximum dynamic-pressure region.

Demonstrate satisfactory canard deployment, launch-escape vehicle turnaround dynamics, and main heat shield forward flight stability prior to launch-escape subsystem (LES) jettison.

Determine the performance of the launch-escape vehicle in the maximum dynamic pressure region.

Demonstrate satisfactory separation of the launch-escape subsystem plus boost-protective cover from the command module.

Demonstrate satisfactory operation and performance of the earth-landing subsystem (LES) using reefed dual-drogues.

Demonstrate satisfactory separation of the launch-escape vehicle from the service module (SM) at an angle of attack.

Demonstrate the structural peformance of the launch-escape subsystem with the canard subsystem.

Demonstrate the structural performance of the boost-protective cover.

Determine the aerodynamic pressure loads on service module during the launch phase. Obtain thermal effects data on the command module during an abort in the maximum dynamic-pressure region.

Demonstrate satisfactory performance of the launch vehicle attitude-control subsystem.

LITTLE JOE A-003
TEST OBJECTIVES
Demonstrate satisfactory launch escape vehicle (LEV) performance at an altitude approximating the upper limit for the canard subsystem.
Demonstrate orientation of the LEV to a main heat shield forward attitude after abort.

LJ-II A003

Determine the damping of LEV oscillations with the canard subsystem deployed.

Demonstrate the separation of the Launch Escape Subsystem (LES) plus Boost Protective Cover (BPC) by the tower jettison motor, and jettison of the forward heat shield by the thrusters.

Determine degradation in window visibility due to rocket motor exhaust products for an abort in the region of abort mode transition altitude.

Determine the physical behavior of the boost protective cover during launch and entry from high altitude.

Obtain data on thermal effects during boost and during impingement of the launch escape motor plumes on the command module and the launch escape tower.

Determine pressures on the command module boost protective cover during launch and high altitude abort.

Demonstrate performance of the earth landing subsystem using the two-point harness attachment for the main parachutes.

Determine vibration and acoustic environment and response of the service module with simulated reaction control subsystem motor quadrants.

LITTLE JOE A-004
TEST OBJECTIVES

Demonstrate satisfactory launch escape vehicle (LEV) performance for an abort in the power-on tumbling boundary region.

Demonstrate the structural integrity of the LEV air-frame structure for an abort in the power-on tumbling boundary region.

LJ-II A004

Demonstrate the capability of the canard subsystem to satisfactorily reorient and stabilize the LEV with the aft heat shield forward after a power-on tumbling abort.

Demonstrate the structural capability of the production boost protective cover to withstand the launch environment.

Determine the static loads on the command module during the launch and the abort sequences.

Determine the dynamic loading of the command module inner structure.

Determine the dynamic loads and the structural response of the service module during launch.

Demonstrate the capability of the command module forward heat shield

thrusters to satisfactorily separate the forward heat shield after the Power has been jettisoned by the Power jettison motor.

Determine the static pressure imposed on the command module by freestream conditions and launch escape subsystem motor plumes during a Power-on tumbling abort.

Obtain data on rendezvous window visibility degradation due to launch escape motor exhaust products for a Power-on tumbling abort.

PAD ABORT TEST I
TEST OBJECTIVES

Test the LES ability to work in emergency before launch while on the pad.

PA 1 & 2
(BP6 & BP23)

PAD ABORT TEST 2
TEST OBJECTIVES

Test the LES ability to work on pad with full BPC, canards, jettisonable apex and dual reefed drogue chutes.

Pad Abort Test I
Nov. 7th 1963

Little Joe II A-001 launch.
May 13th 1964

SA-6 Launch May 28th 1964. The first space
launch of a Saturn/Apollo configuration (above)

Little Joe II A-002 launch.
December 8th 1964

SA-7 Launch Sept. 18th 1964
carrying "Boilerplate 15"
Apollo to orbit

SA-8 Launch May 25th 1965 took
Apollo BP-26 and large Pegasus B
satellite into orbit.

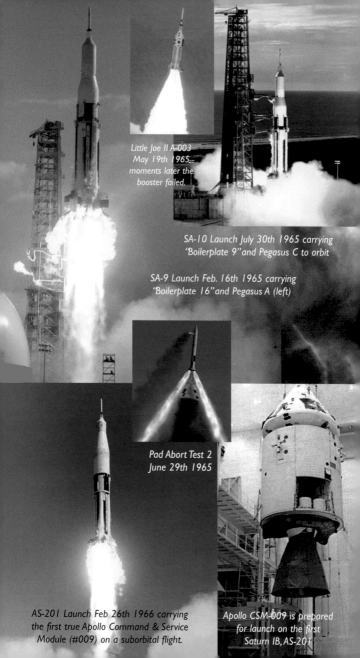

Little Joe II A-003
May 19th 1965...
moments later the
booster failed.

SA-10 Launch July 30th 1965 carrying
"Boilerplate 9" and Pegasus C to orbit

SA-9 Launch Feb. 16th 1965 carrying
"Boilerplate 16" and Pegasus A (left)

Pad Abort Test 2
June 29th 1965

AS-201 Launch Feb 26th 1966 carrying
the first true Apollo Command & Service
Module (#009) on a suborbital flight.

Apollo CSM-009 is prepared
for launch on the first
Saturn IB, AS-201

Little Joe II A-004 begins to disintegrate moments before the ignition of the LES. January 20th 1966

The second launch of a Saturn IB was AS-203 which conducted tests on the fuel. No Apollo was attached for the launch on July 5th 1966. This was the first orbital S-IB flight (right)

AS-202 Launch August 25th 1966 carrying CSM 011 on a suborbital flight to test the heat shield. (left)

The Apollo 204 Command/Service module is prepared for launch (right). This became known as Apollo 1 and was never launched into space.

A beautiful night time shot of the AS-500F Facilities Test Vehicle. This Saturn V replica never flew but was used for ground testing in 1966.

The mighty Saturn V finally takes flight. carrying the Lunar Test Article LTA-10R and CSM-017. November 9th 1967

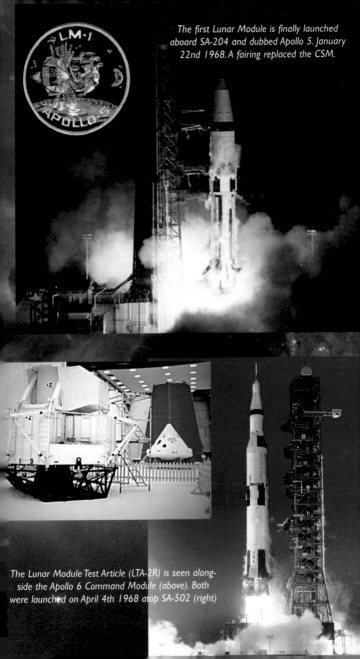

The first Lunar Module is finally launched aboard SA-204 and dubbed Apollo 5. January 22nd 1968. A fairing replaced the CSM.

The Lunar Module Test Article (LTA-2R) is seen alongside the Apollo 6 Command Module (above). Both were launched on April 4th 1968 atop SA-502 (right)

Edward White, Virgil "Gus" Grissom and Roger Chaffee (l to r above) were to be the first to fly Apollo. All three were killed in a launch pad fire on January 27th 1967.

Virgil "Gus" Grissom was one of America's most respected astronauts. He is seen here wearing his Apollo helmet and suit during training for the AS-204 (Apollo 1) mission in 1966.

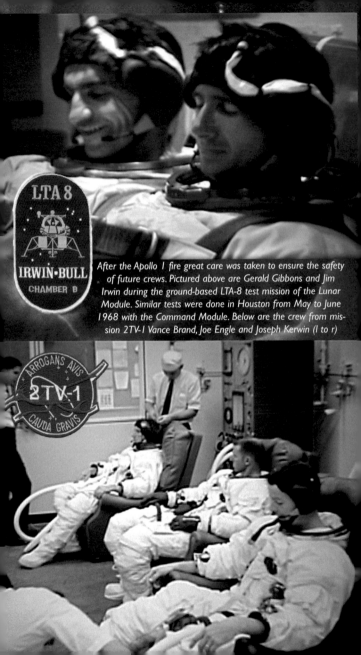

After the Apollo I fire great care was taken to ensure the safety of future crews. Pictured above are Gerald Gibbons and Jim Irwin during the ground-based LTA-8 test mission of the Lunar Module. Similar tests were done in Houston from May to June 1968 with the Command Module. Below are the crew from mission 2TV-1 Vance Brand, Joe Engle and Joseph Kerwin (l to r)

SCHIRRA · EISELE · CUNNINGHAM

VII

USA

The first manned Apollo launch was Apollo 7. Here it is pictured moments after launch from Pad 34 on October 11th 1968.

COMPLEX 34
SATURN I·B Launch Site

Almost two years after the Apollo 1 fire Donn Eisele, Wally Schirra and Walt Cunningham (top) were launched aboard the new 'Block II' Apollo space-craft. It was called Apollo 7 and would be the first manned flight of an Apollo. The Saturn IVB stage did not carry a Lunar Module but had a target for docking practice. (above).

This spectacular picture shows the Apollo 7 S-IVB stage floating above the distinctive coastline of Cape Canaveral in Florida after separating with the Apollo CSM.

Saturn/Apollo vehicle 503 was the first fully operational manned
Saturn V lunar rocket. Here it is awaiting launch in 1968.

The die is finally cast on December 21st 1968 when NASA's gigantic lunar rocket the Saturn V leaves for its first lunar voyage. The launch of Apollo 8 (left).

BORMAN LOVELL ANDERS

The crew of Apollo 8 William Anders, James Lovell and Frank Borman (l to r) were selected to be the first humans to leave Earth orbit.

The crew of Apollo 8 still have fun arguing over who took this most famous photograph of Earth-rise above the lunar horizon. Perhaps one of the most important pictures ever taken. It is worth noting that very similar pictures had been taken earlier by robotic probes but none had the impact of this beautiful color image which was accompanied by eye-witness testimony to an event never before witnessed by human eyes.

The first lunar voyagers splashed down in the Pacific Ocean exactly as predicted by Jules Verne 103 years earlier. Inset is the Apollo 8 command module as it is hauled aboard the recovery ship.

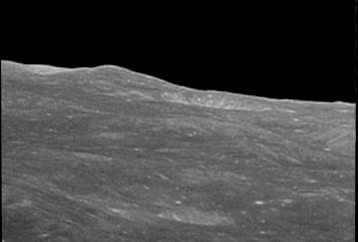

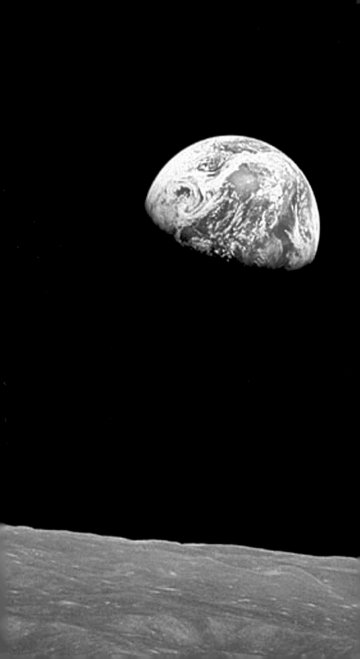

Apollo 9 CSM nicknamed "Gumdrop" this was the first good picture of an Apollo in space. (below)

SA-504 roars off the pad. March 3rd 1969

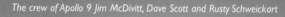

The crew of Apollo 9 Jim McDivitt, Dave Scott and Rusty Schweickart

Two extraordinary pictures of the first complete
Lunar Module "Spider". Above, still inside the
S-IVB and below in free flight.

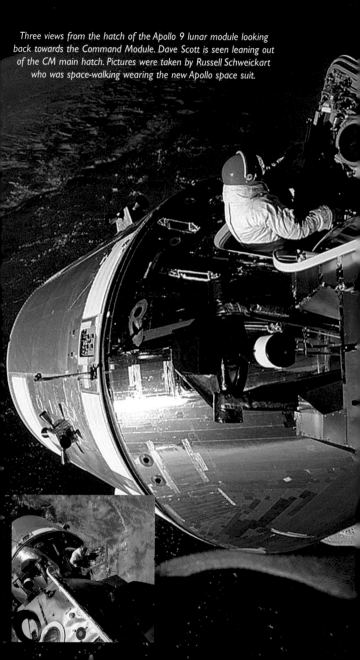

Three views from the hatch of the Apollo 9 lunar module looking back towards the Command Module. Dave Scott is seen leaning out of the CM main hatch. Pictures were taken by Russell Schweickart who was space-walking wearing the new Apollo space suit.

Apollo 10 leaves the Cape on May 18th 1969, just eight years after President Kennedy's famous speech.

The Apollo 10 Lunar Module, nicknamed 'Snoopy', is raised in preparation for installation in the Saturn V. (inset below)

The crew of Apollo 10, Eugene Cernan, Thomas Stafford and John Young (l to r) were all veteran astronauts. They took the final Apollo test flight to within 47,000 feet of the moon. The pictures taken, such as seen here, were spectacular.

The ascent stage of "Snoopy" is seen here just prior to docking with the Command Module "Charlie Brown" (opposite).

The crew of Apollo 10 safely aboard the recovery vessel (right). Apollo's test program was concluded and a trip to the moon was now imminent.